Lecture Notes in Computer Science 15515

Founding Editors

Gerhard Goos
Juris Hartmanis

Editorial Board Members

Elisa Bertino, *Purdue University, West Lafayette, IN, USA*
Wen Gao, *Peking University, Beijing, China*
Bernhard Steffen, *TU Dortmund University, Dortmund, Germany*
Moti Yung, *Columbia University, New York, NY, USA*

The series Lecture Notes in Computer Science (LNCS), including its subseries Lecture Notes in Artificial Intelligence (LNAI) and Lecture Notes in Bioinformatics (LNBI), has established itself as a medium for the publication of new developments in computer science and information technology research, teaching, and education.

LNCS enjoys close cooperation with the computer science R & D community, the series counts many renowned academics among its volume editors and paper authors, and collaborates with prestigious societies. Its mission is to serve this international community by providing an invaluable service, mainly focused on the publication of conference and workshop proceedings and postproceedings. LNCS commenced publication in 1973.

Natasha Lepore · Marius George Linguraru
Editors

Low Field Pediatric Brain Magnetic Resonance Image Segmentation and Quality Assurance

First MICCAI Challenge, LISA 2024
Held in Conjunction with MICCAI 2024
Marrakesh, Morocco, October 10, 2024
Proceedings

Editors
Natasha Lepore
Children's Hospital Los Angeles
Los Angeles, CA, USA

Marius George Linguraru
Children's National Hospital
Washington, DC, USA

ISSN 0302-9743 ISSN 1611-3349 (electronic)
Lecture Notes in Computer Science
ISBN 978-3-031-83010-5 ISBN 978-3-031-83008-2 (eBook)
https://doi.org/10.1007/978-3-031-83008-2

© The Editor(s) (if applicable) and The Author(s) 2025. This book is an open access publication.

Open Access This book is licensed under the terms of the Creative Commons Attribution 4.0 International License (http://creativecommons.org/licenses/by/4.0/), which permits use, sharing, adaptation, distribution and reproduction in any medium or format, as long as you give appropriate credit to the original author(s) and the source, provide a link to the Creative Commons license and indicate if changes were made.
The images or other third party material in this book are included in the book's Creative Commons license, unless indicated otherwise in a credit line to the material. If material is not included in the book's Creative Commons license and your intended use is not permitted by statutory regulation or exceeds the permitted use, you will need to obtain permission directly from the copyright holder.
The use of general descriptive names, registered names, trademarks, service marks, etc. in this publication does not imply, even in the absence of a specific statement, that such names are exempt from the relevant protective laws and regulations and therefore free for general use.
The publisher, the authors and the editors are safe to assume that the advice and information in this book are believed to be true and accurate at the date of publication. Neither the publisher nor the authors or the editors give a warranty, expressed or implied, with respect to the material contained herein or for any errors or omissions that may have been made. The publisher remains neutral with regard to jurisdictional claims in published maps and institutional affiliations.

This Springer imprint is published by the registered company Springer Nature Switzerland AG
The registered company address is: Gewerbestrasse 11, 6330 Cham, Switzerland

If disposing of this product, please recycle the paper.

Preface

Portable ultra-low field (uLF, i.e., 0.064T) magnetic resonance imaging (MRI) offers a lower-cost point-of-care solution to scarce radiological alternatives in resource-limited regions. However, uLF is an emerging technology that requires training for MRI system operators and radiologists in addition to the decrease in image quality compared to high-field MRI. Therefore, automatic methods that confirm image acquisition of appropriate quality for diagnosis and segment and measure critical anatomical structures have the potential to increase the usability and diagnostic quality of uLF MRI.

To support this global effort, we organized the **Low-field pediatric brain magnetic resonance Image Segmentation and quality Assurance (LISA)** Challenge in conjunction with the Medical Image Computing and Computer Assisted Intervention Conference (MICCAI) in Marrakesh, Morocco on October 10, 2024. The LISA Challenge was a benchmarking event for the development and evaluation of automatic image analysis and machine learning algorithms to address two tasks for uLF MRI: Task 1 - automated image quality assurance and Task 2 - automated hippocampal segmentation. With the support of data contributor teams led by Kirsty Donald from the University of Cape Town, South Africa, Sadia Parkar, Sidra Kaleem and Salman Osmani of Aga Khan University, Pakistan and Victoria Nankabirwa of Makerere University, Uganda, we manually curated, rated and released 648 uLF MR images across 7 artifact domains for Task 1. For Task 2, we manually segmented and released bilateral hippocampi across 99 matching high-field images and registered them down to uLF space. All data was made freely available to registered participants of the LISA Challenge.

The LISA challenge welcomed 36 participant teams from 9 countries across 4 continents. All submissions were peer-reviewed through Microsoft's Conference Management Toolkit using a single-blinded process. In these proceedings, we present the work from the three top-performing teams for both tasks, presented as 5 full papers; 2 teams were top performers in both tasks. An additional 6th paper was presented outside of the competition and ranking by the organizing team and discusses baseline approaches for resolving the challenge tasks. The winning team of Task 1 for quality assessment was from the Department of Computational and Data Sciences, Indian Institute of Science, Bangalore, India. The winning team of Task 2 on hippocampal segmentation was from Asan Medical Institute of Convergence Science and Technology, University of Ulsan College of Medicine, Asan Medical Center, Seoul, South Korea.

We would like to thank all the teams who contributed to the challenge, the MICCAI 2024 Conference and especially the challenge chairs Shadi Albarqouni, Yunusa Mohammed and Spyridon Bakas for hosting us, and the Bill and Melinda Gates Foundation for providing the financial support and resources to organize the LISA Challenge. We also thank all the organizers and data contributors who made this event a community success. In addition to the two of us, these include Sean Deoni, Austin Tapp, Jeffrey Tanedo, Rahimeh Rouhi, Shreyash Zanjal, Eryn Perry, Steve Williams, Kirsty Donald, Victoria Nankabirwa, Sadia Parkar and Salman Osmani. This work was supported by the

Bill & Melinda Gates Foundation under investments INV-005798, INV-047887, INV-018164, INV-004939 and INV-023509, as well as the Wellcome Leap 1kD programme (The First 1000 Days; 222076/Z/20/Z).

Ultra-low-field brain imaging has the potential to become a transformative tool for both clinical and research applications, particularly in limited-resource settings where the need for medical imaging democratization is the most critical. It is our hope that the methods developed as part of the LISA Challenge will help pave the way toward broader use of this new and impactful technology. See you at LISA 2025!

November 2024

Natasha Lepore
Marius George Linguraru

Organization

Program Committee Chairs

Natasha Lepore — Children's Hospital Los Angeles, USA
Marius George Linguraru — Children's National Hospital, USA

Program Committee

Sean Deoni — Bill and Melinda Gates Foundation, USA
Kirsty Donald — University of Cape Town, South Africa
Sidra Kaleem — Aga Khan University, Pakistan
Victoria Nankabirwa — Makerere University, Uganda
Salman Osmani — Aga Khan University, Pakistan
Sadia Parkar — Aga Khan University, Pakistan
Rahimeh Rouhi — Children's Hospital Los Angeles, USA
Austin Tapp — Children's National Hospital, USA
Jeffrey Tanedo — Children's Hospital Los Angeles, USA
Steve Williams — King's College London, UK
Shreyash Zanjal — Children's Hospital Los Angeles, USA

Additional Reviewers

Athelia Paulli — Children's Hospital Los Angeles, USA
Eryn Perry — Children's Hospital Los Angeles, USA

Contents

Task 1 - Automatic Ultra-Low Field MR Image Quality Assessment

Multi-Label MambaOut for Quality Assessment of Low-Field Pediatric
Brain MR Images 3
   Yueyue Zhu, Haotian Jiang, Rongqing Cai, and Geng Chen

Task 2 - Automatic Hippocampal Segmentation from Ultra-Low Field MRI

Bilateral Hippocampi Segmentation in Low Field MRIs Using Mutual
Feature Learning via Dual-Views 15
   Himashi Peiris and Zhaolin Chen

Infant Hippocampal Segmentation in Ultra-Low-Field MRI Using External
Datasets with Diverse Field Strengths 28
   Weichen Zhou, Jingyu Li, Xi Wang, Yuwan Wang, and Mengye Lyu

Task 1 and Task 2 Combined

Automated Quality Assessment Using Appearance-Based Simulations
and Hippocampus Segmentation on Low-Field Paediatric Brain MR Images 41
   Vaanathi Sundaresan and Nicola K Dinsdale

Axis-Guided Quality Assessment and Multi-label Hippocampal
and Ventricular Segmentation in Low-Resolution Pediatric Brain MRI 53
   Hyunwook Kim, Jinew Seo, Seiyoung Ryu, Joon hyung Park, Sungchul On, and Jinwha Choi

Quality Assurance and Hippocampal Segmentation on Low-Field Pediatric
Magnetic Resonance Images 63
   Austin Tapp, Rahimeh Rouhi, Jeffrey Tanedo, Shreyash Zanjal, Sean Deoni, Marius George Linguraru, and Natasha Lepore

Author Index 77

Task 1 - Automatic Ultra-Low Field MR Image Quality Assessment

Multi-Label MambaOut for Quality Assessment of Low-Field Pediatric Brain MR Images

Yueyue Zhu, Haotian Jiang, Rongqing Cai, and Geng Chen[✉]

National Engineering Laboratory for Integrated Aero-Space-Ground-Ocean Big Data Application Technology, School of Computer Science and Engineering, Northwestern Polytechnical University, Xi'an, China
geng.chen.cs@gmail.com

Abstract. Magnetic Resonance Imaging (MRI) can be utilized to study the structure of pediatric brains non-invasively. In practice, low-field MRI scanners are widely adopted for pediatric brain imaging. However, the corresponding acquired MRI data usually suffers from severe artifacts, such as noise and motion. Therefore, an effective Quality Assessment (QA) method is essential. To this end, we design a Multi-Label MambaOut (MLMambaOut) model for the low-field pediatric brain MRI QA challenge. Specifically, we view this challenge as a multi-label classification task, utilizing four stages of gated convolution neural network blocks and ML-Decoder to finish the classification with class balance loss. Furthermore, we explore the performance of Mamba and some advanced models for this challenge. We performed extensive experiments on the challenge data, which is low-field and corrupted with seven kinds of artifacts. The results show that our MLMambaOut achieves superior classification results compared with other methods.

Keywords: Low-field pediatric brain · Quality assessment · Multi-label classification

1 Introduction

Magnetic Resonance Imaging (MRI) is a vital tool in the examination of pediatric brains. However, due to uncontrolled motion, the acquired pediatric MRI data are usually accompanied by motion artifacts, noise, etc. Therefore, Quality Assessment (QA) of the acquired data ensures that it meets the specific criteria for clinical diagnosis and neuroscience studies. However, only low-field 0.064T MRI scans can currently be collected in some developing countries. Although such data are acquired at a low cost, the extremely low resolution creates a significant challenge for QA assessment. Therefore, it is essential to develop a QA method for evaluating low-field 0.064T MRI data in pediatrics.

Equal contribution—Y. Zhu and H. Jiang.
This work was supported in part by the National Natural Science Foundation of China (No. 61540047) and the Practice and Innovation Funds for Graduate Students of Northwestern Polytechnical University (No. PF2024012).

With the development of deep learning, some methods based on Convolutional Neural Networks (CNNs) have been proposed. For instance, Xu et al. [11] proposed a semi-supervised learning method that improved the accuracy of fetal brain MRI images for artifact quality assessment by introducing region of interest (ROIs) consistency in the fetal brain. Masoudi et al. [8] improved the accuracy of image QA using generative adversarial networks in MRI image QA of prostate cancer. Liu et al. [5] designed residual blocks constructed by deep separable convolution and non-local mean operations to significantly improve the automation and accuracy of the evaluation through hierarchical processing and efficient use of limited labeled data. Kastryul et al. [3] evaluated and provided in-depth analyses and valuable insights into distorted images for MRI. Sanchez et al. [10] applied existing methods to QA and quality of fetal brain control by automating regression and classification tasks. However, these methods are based on CNNs and are unable to capture long-range dependencies. The transformer, which can compensate for this deficiency, provides an effective solution for QA tasks. For example, Zhang et al. [13] presented a deep learning framework that handles both brain extraction and image quality assessment by incorporating the transformer architecture into both the feature extraction module and the segmentation header. Most of the aforementioned methods simply evaluate a specific artifact category and are trained and validated on high-quality adult data. The Low-field Pediatric Brain Magnetic Resonance Image Segmentation and Quality Assurance (LISA) challenge aims to identify seven categories of artifacts, including "Noise", "Zipper", "Positioning", "Banding", "Motion", "Contrast", and "Distortion". Additionally, datasets of LISA have a low resolution with lowfield 0.064T MRI data in infants. This is a difficult task and the above methods are not directly applicable to this challenge.

Currently, a Transformer, which can capture global information, is a mainstream model for image classification tasks, however, it is not linear in complexity and requires higher memory consumption. Compared with the Transformer, the Mamba model [1] emerged as an RNN-like State-Space Model (SSM) with linear complexity, but its performance in visual tasks is unsatisfactory. More recently, MambaOut [12] was proposed and points out that, for the ImageNet image classification task, SSM is not necessary and can lead to performance degradation.

To this end, we view the challenge QA task as a multi-label classification one and design an effective MambaOut-based classification model for this task. Our model, called Multi-Label MambaOut (MLMambaOut), utilizes the MambaOut framework with ML-Decoder [9] as the output layer and multi-label classification task loss. Meanwhile, due to the imbalance of the dataset classes, we further introduce class-balanced loss to solve this problem. We performed extensive experiments to verify the effectiveness of our MLMambaOut. Compared with cutting-edge classification models, MLMambaOut exhibits superior performance in identifying various artifacts. Our MLMambaOut code is publicly available at https://github.com/zyyNUPU/MLMambaOut.

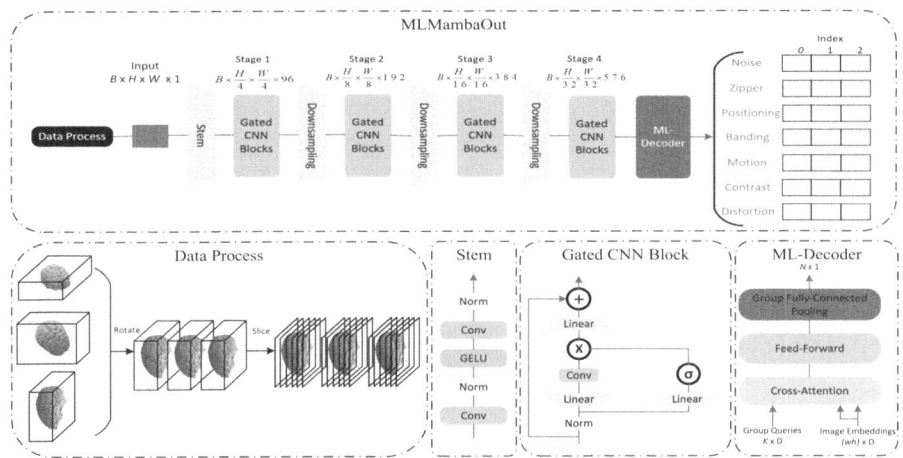

Fig. 1. The overall architecture of MLMambaOut for multi-label classification. Our MLMambaOut consists of four stages, each of which is with a corresponding number of GCBs: 3, 4, 27, and 3.

2 Method

As shown in Fig. 1, the input data $M \in \mathbb{R}^{B \times H \times W \times C_0}$ is fed into MLMambaOut, which includes a stem layer and four stages of Gated CNN Blocks (GCBs). Next, MLMambaOut utilizes ML-Decoder as classification heads to classify "Noise", "Zipper", "Positioning", "Banding", "Motion", "Contrast", and "Distortion". We also introduce class-balanced loss to balance the number of unbalanced classes.

2.1 MLMambaOut

Firstly, the M through the stem layer, including two CNNs, two norm layers, and an activation function, to $M_1 \in \mathbb{R}^{B \times H \times W \times C_1}$, where B, H, W, C_1 represent the batch size, height, width, and number of channels, respectively. Next, we extract any pixel $G \in \mathbb{R}^{B \times C_1}$ from M_1, and the same goes for other pixels. Given the pixel as an input G, the main architecture of GCB is formulated as:

$$G^{'} = \text{Norm}(G), \tag{1}$$

$$R = (\text{Conv}(G^{'} W_1) \odot \text{GELU}(G^{'} W_2))W_3 + G, \tag{2}$$

where $\text{Norm}(\cdot)$ represents normalization, R means the output of GCB, $\text{Conv}(\cdot)$ refers to the convolution operation and $\odot$ stands for Hadamard product; $W_1 \in \mathbb{R}^{C_1 \times rC_1}$, $W_2 \in \mathbb{R}^{C_1 \times rC_1}$ and $W_3 \in \mathbb{R}^{rC_1 \times C_1}$ are learnable parameter with MLP expansion r.

Table 1. Dataset Information.

Labels	Count-0	Count-1	Count-2	Ratio-0	Ratio-1	Ratio-2
Noise	369	49	14	0.8542	0.1134	0.0324
Zipper	347	76	9	0.8032	0.1759	0.0209
Positioning	389	40	3	0.9005	0.0926	0.0069
Banding	417	14	1	0.9653	0.0324	0.0023
Motion	331	72	29	0.7662	0.1667	0.0671
Contrast	317	108	7	0.7338	0.2500	0.0162
Distortion	378	47	7	0.8750	0.1088	0.0162

2.2 Cross Entropy Loss

In this task, the cross entropy loss function is used to measure the difference between prediction and ground truth. Given the true distribution p and the predicted distribution q, the cross entropy loss can be defined as follows:

$$\mathcal{L}(p, q) = -\sum_{i=1}^{C} p_i \log(q_i), \tag{3}$$

where C is the number of classes. p_i is the probability of class i in the true distribution p. In a classification setting, p_i is typically a one-hot encoded vector where the true class has a probability of one, and all other classes have a probability of zero. q_i is the predicted probability of class i from the model. It is the output of the model and is usually obtained by applying the softmax function to the raw output logits. The softmax function converts the logits z_i into probabilities q_i:

$$q_i = \frac{\exp(z_i)}{\sum_{j=1}^{C} \exp(z_j)}, \tag{4}$$

2.3 Class Balance Loss

In practice, the number of artifact-free images is much larger than that of artifact-corrupted ones, raising severe class imbalance issues. To address this issue, we utilize a class balancing loss [4] to train our model.

For notational convenience, we define p_t:

$$p_t = \begin{cases} q_0 & \text{if } y = 0 \\ q_1 & \text{if } y = 1 \\ q_2 & \text{if } y = 2 \end{cases}, \tag{5}$$

where p_t is the probability of the model predicting a true category, $y \in \{0, 1, 2\}$ specifies the ground-truth class, q_0, q_1, q_2 represent the probability of the corresponding class.

The class balance loss is defined as follows:

$$B_{loss}(p_t) = -\alpha (1-p_t)^\gamma \log(p_t), \qquad (6)$$

where α is a weighting factor and the γ is a tunable parameter. $(1-p_t)^\gamma$ can automatically reduce the contribution of simple classes, thus resolving the problem of class balancing.

Table 2. Quantitative comparison of MLMambaOut and other methods. The best of these results are shown in red and the second best in blue.

Methods	Noise	Zipper	Positioning	Banding	Motion	Contrast	Distortion	Mean
Vision Mamba	0.8575	0.8172	0.8498	0.9366	0.7286	0.6978	0.7614	0.8070
MobileNet	0.9422	0.8638	0.8667	0.9526	0.8256	0.8480	0.7349	0.8620
Resnet50	0.9456	0.8861	0.9349	0.9313	0.8567	0.8994	0.8615	0.9022
Swin-Transformer	0.9581	0.9582	0.9615	0.9829	0.8448	0.8783	0.9123	0.9280
MLMambaOut	0.9513	0.9575	0.9560	0.9808	0.8766	0.8882	0.9234	0.9334

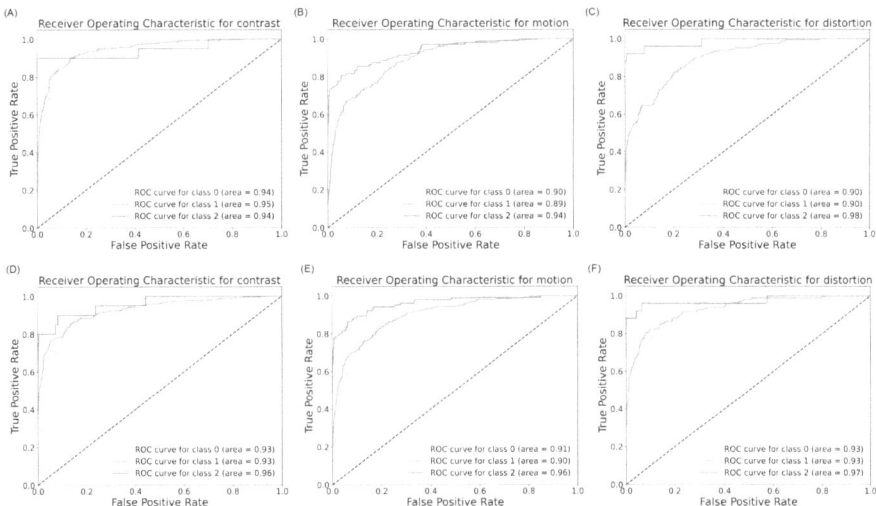

Fig. 2. We show the ROC curves for the MLMambaOut model with and without the class balance loss, where A, B, and C do not use class balance loss, and the rest use.

3 Results

3.1 Dataset Preparation

The LISA challenge provides 432 training data of the 0.064T MRI from three different scanners. The details of the data are shown in Table 1. We extracted

2D high-resolution slices from the MR images of training subjects. Additionally, we used data augmentations, including flipping the image on two axes with a random probability, rotating it within a range of -10 to 10°C, and finally adding a certain amount of random noise. The intensity of the noise is controlled by a small noise factor, which is 0.01 multiplied by a randomly generated value.

3.2 Implementation Details and Evaluation Metrics

The implementation of our network was based on PyTorch. We conducted the training process with an ADAM optimizer on a GTX 3090 GPU with 24GB memory. The learning rate is set to 0.0001. The model's parameter selection is given by depths=[3, 4, 27, 3], which corresponds to the four stages of Gated CNN Blocks (GCBs). The dimensions were chosen as [96, 192, 384, 576], representing the number of channels used in each of these four stages. During the training process, α was set to 4.0 and γ was set to 2.0. We compared the methods including MobileNet [2], Resnet50 [7], Swin-Transformer [6], and Vision Mamba [14].

3.3 Results

The classification results of the different methods are shown in Table 2. Our MLMambaOut outperforms other methods in terms of average accuracy. Although MLMambaOut does not show the best performance in some classes, it provides high accuracy in all classes and the best overall performance. It is worth noting that the popular Mamba model is unable to provide satisfactory performance. In addition, Resnet50 and Swin-Transformer provide relatively satisfactory results, but are unable to deal with some artifacts, such as "Motion". Additionally, it can be seen that most of the models classify "Motion", "Contrast", and "Distortion" relatively poorly. This phenomenon needs to be taken care of in the future. We further present the ROC curves of the MLMambaOut models with and without the class imbalance loss. As shown in Fig. 2, it can be seen that the use of the class balancing loss appropriately mitigates this problem. For the "Motion" and "Contrast" labels, MLMambaOut with balance loss improves the 2% performance at class 2. For the sum of Area Under the Curve (AUC) of classes 0, 1, and 2, MLMambaOut with balance loss shows performance improvements of 4% and 5% on the "Motion" and "Distortion" labels, respectively, but decreased by 1% on the "Contrast" label. To further validate the classification performance of MLMambaOut, we conducted a t-SNE test, as shown in Fig. 3. t-SNE is a dimensionality reduction technique used to visualize high-dimensional data, enabling intuitive analysis of model classification performance by revealing overlaps and separations between different categories. It also aids in identifying outliers, facilitating the evaluation and optimization of feature learning efficacy. From a clustering perspective, Fig. 3 clearly and distinctly illustrates three clusters, indicating that MLMambaOut can effectively categorize each label. Additionally, considering the number of discrete points, it is evident that the number of clustered points greatly exceeds that of the discrete points across all categories, which demonstrates the strong generalization capability of MLMambaOut. Finally, regarding the overlap between classes, Fig. 3 shows that overlap only occurs at the boundaries where two classes meet, with no overlap in deeper regions. This indirectly reflects the robust classification performance of MLMambaOut.

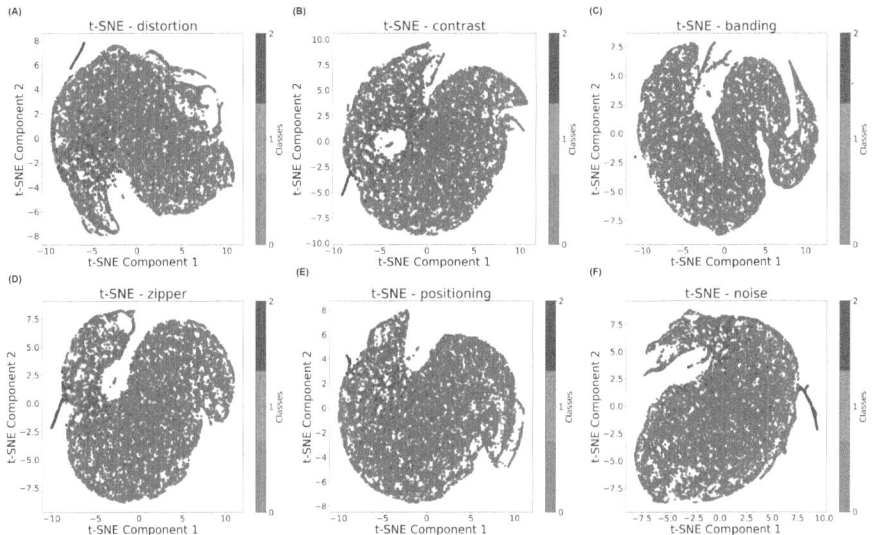

Fig. 3. We show the t-SNE on the MLMambaOut model for six artifact categories, including the banding, zipper, positioning, noise, contrast, and distortion.

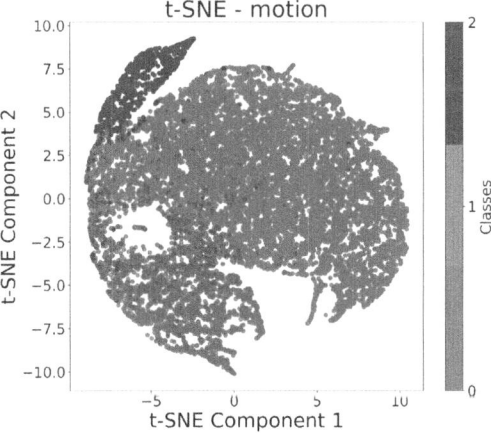

Fig. 4. We show the t-SNE on the MLMambaOut model for "Motion" category.

4 Discussion

Our experiments show that MLMambaOut is effective in classifying most artifacts. However, the final accuracy reveals that the results for the "Motion" artifact are lower compared with other labels. This issue is illustrated in Fig. 4, where the blue region associated with "Motion" is larger than any blue region in Fig. 3 for other labels. As a result, the overlap area between classes for the "Motion" label, as shown in Fig. 4, is larger compared with the corresponding areas for the other six labels in Fig. 3, leading to a reduction in performance of MLMambaOut. To address this issue, data augmentation will be performed specifically for the "Motion" label to increase the quantity of its features. Additionally, since MLMambaOut employs ML-Decoder as output layers, which is beneficial for learning from complex data such as MRI, we plan to replace the ML-Decoder with a more suitable output layer in future work.

5 Conclusion

We proposed MLMambaOut, a multi-label classification model based on MambaOut for the QA of low-field 0.064T pediatric MRI data. Our MLMambaOut achieves promising classification performance and outperforms existing models. Through extensive experimentation, we find that viewing the QA task as a multi-label classification one for MR images is an effective solution. In the future, we will explore further improving the performance with more effective data augmentation for the classes that are difficult to classify.

References

1. Gu, A., Dao, T.: Mamba: linear-time sequence modeling with selective state spaces. arXiv preprint arXiv:2312.00752 (2023)
2. Howard, A.G., et al.: Mobilenets: efficient convolutional neural networks for mobile vision applications. arXiv preprint arXiv:1704.04861 (2017)
3. Kastryulin, S., Zakirov, J., Pezzotti, N., Dylov, D.V.: Image quality assessment for magnetic resonance imaging. IEEE Access **11**, 14154–14168 (2023)
4. Li, X., et al.: Generalized focal loss: learning qualified and distributed bounding boxes for dense object detection. In: Advances in Neural Information Processing Systems, vol. 33, pp. 21002–21012 (2020)
5. Liu, S., Thung, K.H., Lin, W., Shen, D., Yap, P.T.: Hierarchical nonlocal residual networks for image quality assessment of pediatric diffusion MRI with limited and noisy annotations. IEEE Trans. Med. Imaging **39**(11), 3691–3702 (2020)
6. Liu, Z., et al.: Swin transformer: hierarchical vision transformer using shifted windows. In: Proceedings of the IEEE/CVF International Conference on Computer Vision, pp. 10012–10022 (2021)
7. Mascarenhas, S., Agarwal, M.: A comparison between vgg16, vgg19 and resnet50 architecture frameworks for image classification. In: 2021 International Conference on Disruptive Technologies for Multi-Disciplinary Research and Applications (CENTCON), vol. 1, pp. 96–99 (2021). https://doi.org/10.1109/CENTCON52345.2021.9687944

8. Masoudi, S., et al.: No-reference image quality assessment of t2-weighted magnetic resonance images in prostate cancer patients. In: 2021 IEEE 18th International Symposium on Biomedical Imaging (ISBI), pp. 1201–1205. IEEE (2021)
9. Ridnik, T., Sharir, G., Ben-Cohen, A., Ben-Baruch, E., Noy, A.: Ml-decoder: scalable and versatile classification head. In: Proceedings of the IEEE/CVF Winter Conference on Applications of Computer Vision, pp. 32–41 (2023)
10. Sanchez, T., Esteban, O., Gomez, Y., Eixarch, E., Cuadra, M.B.: FetMRQC: automated quality control for fetal brain MRI. In: Link-Sourani, D., Abaci Turk, E., Macgowan, C., Hutter, J., Melbourne, A., Licandro, R. (eds.) Perinatal, Preterm and Paediatric Image Analysis: 8th International Workshop, PIPPI 2023, Held in Conjunction with MICCAI 2023, Vancouver, BC, Canada, October 12, 2023, Proceedings, pp. 3–16. Springer Nature Switzerland, Cham (2023). https://doi.org/10.1007/978-3-031-45544-5_1
11. Xu, E., et al.: Semi-supervised learning for fetal brain MRI quality assessment with ROI consistency. In: Martel, A.L., et al. (eds.) MICCAI 2020. LNCS, vol. 12266, pp. 386–395. Springer, Cham (2020). https://doi.org/10.1007/978-3-030-59725-2_37
12. Yu, W., Wang, X.: Mambaout: do we really need mamba for vision? arXiv preprint arXiv:2405.07992 (2024)
13. Zhang, W., et al.: A joint brain extraction and image quality assessment framework for fetal brain MRI slices. Neuroimage **290**, 120560 (2024)
14. Zhu, L., Liao, B., Zhang, Q., Wang, X., Liu, W., Wang, X.: Vision mamba: efficient visual representation learning with bidirectional state space model. arXiv preprint arXiv:2401.09417 (2024)

Open Access This chapter is licensed under the terms of the Creative Commons Attribution 4.0 International License (http://creativecommons.org/licenses/by/4.0/), which permits use, sharing, adaptation, distribution and reproduction in any medium or format, as long as you give appropriate credit to the original author(s) and the source, provide a link to the Creative Commons license and indicate if changes were made.

The images or other third party material in this chapter are included in the chapter's Creative Commons license, unless indicated otherwise in a credit line to the material. If material is not included in the chapter's Creative Commons license and your intended use is not permitted by statutory regulation or exceeds the permitted use, you will need to obtain permission directly from the copyright holder.

Task 2 - Automatic Hippocampal Segmentation from Ultra-Low Field MRI

Bilateral Hippocampi Segmentation in Low Field MRIs Using Mutual Feature Learning via Dual-Views

Himashi Peiris[1,2(✉)] and Zhaolin Chen[1,2]

[1] Department of Data Science and AI, Faculty of IT, Monash University, Melbourne, Australia
{Himashi.Peiris,Zhaolin.Chen}@monash.edu
[2] Monash Biomedical Imaging (MBI), Monash University, Melbourne, Australia

Abstract. Accurate hippocampus segmentation in brain MRI is critical for studying cognitive and memory functions and diagnosing neurodevelopmental disorders. While high-field MRIs provide detailed imaging, low-field MRIs are more accessible and cost-effective, which eliminates the need for sedation in children, though they often suffer from lower image quality. In this paper, we present a novel deep-learning approach for the automatic segmentation of bilateral hippocampi in low-field MRIs. Extending recent advancements in infant brain segmentation to underserved communities through the use of low-field MRIs ensures broader access to essential diagnostic tools, thereby supporting better healthcare outcomes for all children. Inspired by our previous work, Co-BioNet, the proposed model employs a dual-view structure to enable mutual feature learning via high-frequency masking, enhancing segmentation accuracy by leveraging complementary information from different perspectives. Extensive experiments demonstrate that our method provides reliable segmentation outcomes for hippocampal analysis in low-resource settings. The code is publicly available at: https://github.com/himashi92/LoFiHippSeg.

Keywords: Hippocampi Segmentation · Low-field MRI · Feature Learning · Dual-view Learning · Frequency Masking

1 Introduction

The hippocampus is a vital subcortical structure in memory formation and cognitive processes. Accurate hippocampus segmentation in MRI scans is essential for studying neurodevelopmental disorders and cognitive impairments. High-field MRIs, with their superior image quality, are typically used for this task, but their high cost and limited availability pose significant barriers, especially in low-resource settings [13]. Low-field MRIs, while more accessible, often produce images with lower resolution and increased noise, making accurate hippocampal segmentation challenging. Recent advancements in deep learning have shown promise in improving medical image segmentation [2,10]. However, existing methods primarily focus on high-field MRIs, leaving a gap in effective techniques for low-field MRI segmentation. Inspired by the mutual feature learning mechanism of Co-BioNet [12], we propose a novel approach that utilizes a dual view structure to enhance segmentation performance in low-field MRIs. By learning complementary features from two different views by associating high-frequency images of the given -field MR via high-frequency masking, our model can effectively capture the complex structures of the bilateral hippocampi. This approach of utilizing high-frequency images alongside the original low-field images demonstrates potential as a valuable tool for improving the usability of low-field images in low-resource settings, as it minimizes the need for extensive external tools and datasets.

This paper presents our dual-view mutual feature learning framework and demonstrates its efficacy through extensive experiments on the LISA 2024 low-field MRI dataset. Our results highlight the potential of this approach in providing accurate and reliable hippocampal segmentation, thereby facilitating better diagnostic and research capabilities in resource-constrained environments (Fig. 1).

2 Dataset

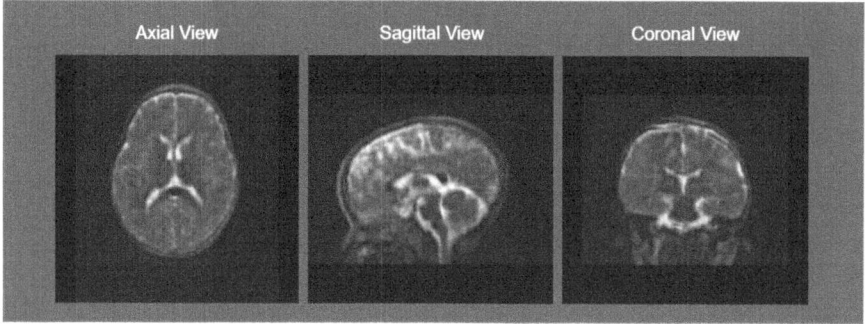

Fig. 1. Sample case from LISA Dataset.

The dataset utilized in this study comprises high-field T2-weighted MRI scans and synchronized low-field Hyperfine scans acquired from institutions in Uganda, South Africa, and the United States [4]. Expert MRI technicians collected the images, ensuring high-quality data. The dataset includes meticulously reviewed bilateral hippocampi segmentations by an expert medical image evaluator, providing a reliable ground truth. The images are available in NIFTI (.nii.gz) format, with low-field images registered to high-field scans through a 9-point linear registration process. Orthogonal low-field images were processed using the ANTs multivariate template construction and aligned with a pediatric T2 template, subsequently coregistered to matching high-field scans using FLIRT from the FSL toolbox [14]. This robust dataset underpins developing and evaluating our deep-learning model for accurate hippocampal segmentation in low-field MRIs.

3 Methodology

3.1 Notations and Problem Formulation

In our paper, we represent vectors and matrices using bold lowercase $\mathbf{x}$ and bold uppercase $\mathbf{X}$, respectively. The norm of a vector is denoted by $\|\cdot\|$, with $\|\mathbf{x}\|_1 = \sum_i |\mathbf{x}[i]|$, where $\mathbf{x}[i]$ signifies the element at position i in $\mathbf{x}$. The inner product between vectors is denoted by $\langle \cdot, \cdot \rangle$, and $\|\mathbf{x}\|_2^2 = \langle \mathbf{x}, \mathbf{x} \rangle$. When norms and inner products are applied to 3D tensors, we assume the tensors are flattened. For instance, for 3D tensors $\mathbf{A}$ and $\mathbf{B}$, $\langle \mathbf{A}, \mathbf{B} \rangle = \sum_{i,j,k} \mathbf{A}[i,j,k]\mathbf{B}[i,j,k]$ and $\|\mathbf{A}\|_1 = \sum_{i,j,k} |\mathbf{A}[i,j,k]|$.

Consider a dataset $\mathcal{X}_1 = \{(\mathbf{X}_i, \mathbf{Y}_i)\}_{i=1}^n$ consisting of n samples, where each sample $(\mathbf{X}_i, \mathbf{Y}_i)$ includes an image $\mathbf{X}_i \in \mathbb{R}^{C \times H \times W \times D}$ and its corresponding ground-truth segmentation mask $\mathbf{Y}_i \in \{0,1\}^{K \times H \times W \times D}$, encoded as a one-hot K-dimensional vector for a K-class problem per voxel. Here, C, H, W, and D denote the number of channels, height, width, and depth of the input medical volume. Similarly, in order to create Dual-Views of the input, we use a high pass filtering method to generate a high frequency of the low-field MR volume, which creates another dataset $\mathcal{X}_2 = \{(\hat{\mathbf{X}}_i, \mathbf{Y}_i)\}_{i=1}^n$ consisting of n samples. The primary objective is to co-learn segmentation models from $\mathcal{D} = \mathcal{X}_1 \cup \mathcal{X}_2$.

3.2 LoFiHippSeg Architecture

Inspired by our previous works [8,12], we propose a Dual-View deep learning architecture named **LoFiHippSeg** to learn features from **Lo**w-**Fi**eld MRIs for **Hipp**ocampi **Seg**mentaion. As shown in the conceptual diagram in Fig. 2, we use two segmentation networks denoted as $\mathcal{F}_1(\cdot)$ and $\mathcal{F}_2(\cdot)$, which creates Dual-Views which mutually learn from complementary features. $\mathcal{X}_1$ dataset and $\mathcal{X}_2$ is used to train two segmentation models, respectively. Considering the computational complexity in our pipeline, we use VNet [7] as the segmentation model for dual-view training. As the critic network, we use a fully convolutional neural network similar to encoder architecture, following recent works [12].

Frequency Masking Module (FMM). Frequency domain analysis is crucial in medical imaging, including MRI reconstruction and image-denoising applications. The low frequencies in an image's Fourier spectrum represent the mean image intensity (DC signal) and the intensities of significant image components. Conversely, high frequencies capture fine details such as edges, boundaries between tissues, and the delicate outlines of structures [5]. Considering this, we used the data augmentation based on the frequency masking approach to create another view of the original low-field MR volume. We employ both the original low-field MR volume and its high-frequency components to train our dual views based on the principle of consensus [1,3,15]. This approach ensures that the complementary information from both views is integrated, enhancing the overall performance of the model. The original low-field MR volume provides essential structural and intensity information, capturing the general anatomy

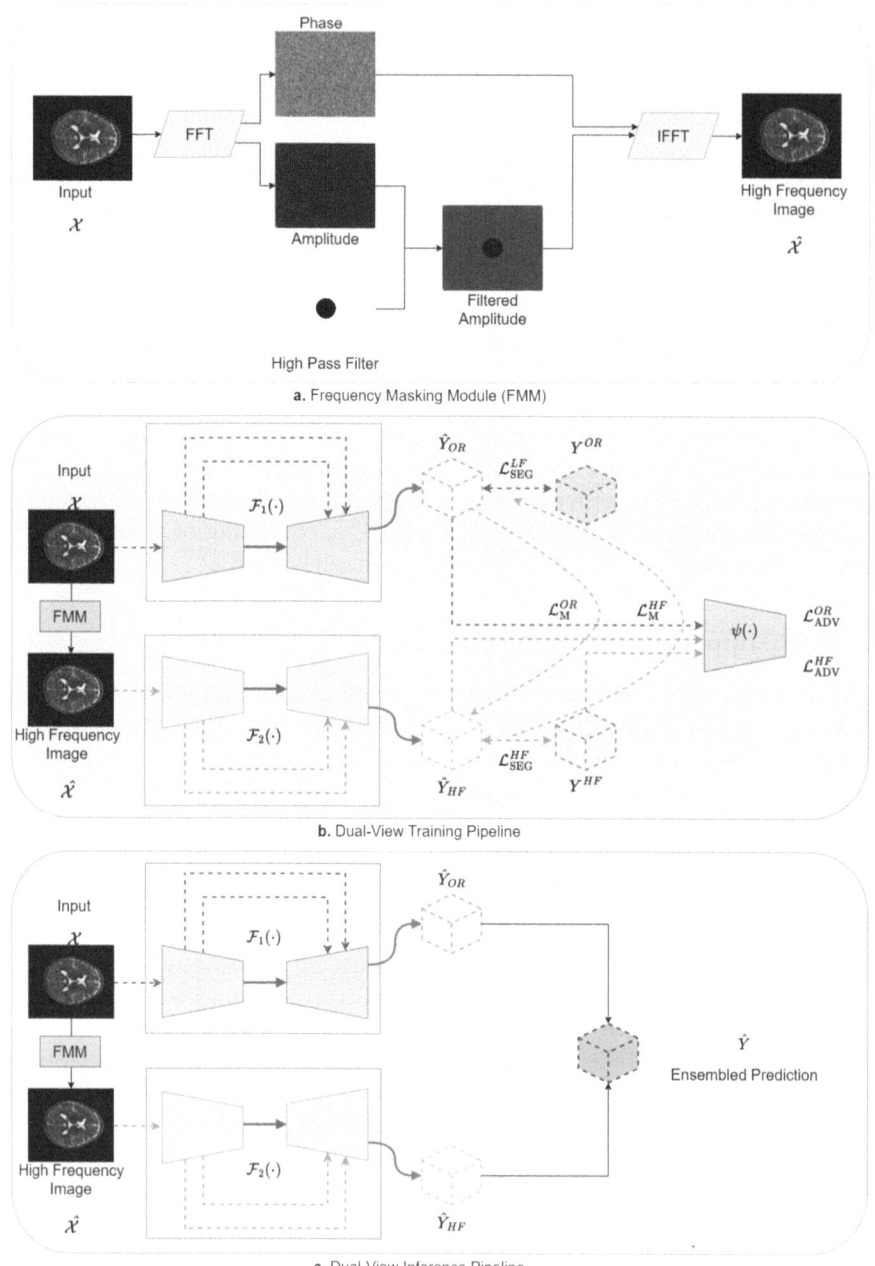

Fig. 2. Overview of Proposed Dual-View Pipeline, **LoFiHippSeg.** Here, $\mathcal{F}_1(\cdot)$ and $\mathcal{F}_2(\cdot)$ are structuraly similar VNet models.

and major tissue contrasts. In contrast, the high-frequency components emphasize fine details such as edges and boundaries, which are crucial for accurately delineating structures. By training with both views, the network can leverage the strengths of each, leading to more robust and precise segmentation. The consensus between these dual views helps reinforce consistent and accurate predictions, ultimately improving the model's reliability and effectiveness in various medical imaging tasks.

Consider a low-field MR volume $\mathbf{X} \in \mathbb{R}^{C \times H \times W \times D}$. To perform frequency masking, we first transform $\mathbf{X}$ to the frequency domain using the Fourier transform $\mathcal{FFT}$.

$$\mathbf{X}_f = \mathcal{FFT}(\mathbf{X}) \tag{1}$$

The Fourier-transformed medical volume $\mathbf{X}_f$ can be decomposed into its amplitude $\mathbf{A}_{\mathbf{X}_f}$ and phase $\mathbf{P}_{\mathbf{X}_f}$ components. Next, we define a high-pass filter $\mathbf{H} \in \{0,1\}^{D \times H \times W}$ that selectively retains high-frequency components. The mask is designed such that it zeros out the low-frequency components and retains the high-frequency ones:

$$\mathbf{A}_{\mathbf{X}_f}^{\text{high}} = \mathbf{A}_{\mathbf{X}_f} \odot \mathbf{H} \tag{2}$$

Here, $\odot$ denotes the element-wise multiplication. The phase remains unchanged: The high-frequency representation $\hat{\mathbf{X}}_{\text{high}}$ in the frequency domain is then:

$$\hat{\mathbf{X}}_{\text{high}} = \mathbf{A}_{\mathbf{X}_f}^{\text{high}} \cdot e^{i\mathbf{P}_{\mathbf{X}_f}} \tag{3}$$

This high-frequency representation is then transformed back to the spatial domain using the inverse Fourier transform $\mathcal{FFT}^{-1}$:

$$\hat{\mathbf{X}} = \mathcal{FFT}^{-1}(\hat{\mathbf{X}}_{\text{high}}) \tag{4}$$

The resulting $\hat{\mathbf{X}}$ captures the high-frequency components of the original image volume $\mathbf{X}$.

3.3 Objective Function

Building on our previous works [8,12], we train each segmentation network (VNet model), by optimizing the following min-max problem:

$$\min_{\theta_i} \max_{\theta_c} \; \mathcal{L}^i(\Theta; \mathcal{D}) \,. \tag{5}$$

Here, Θ includes all the networks' parameters, i.e., $\theta_1, \theta_2, \theta_c$. The min-max problem in Equation Eq. 5 aims to determine whether the prediction masks generated by the segmentation networks belong to the same distribution as the ground truth or if they deviate from it. Here, both original low field MRI and High-frequency Images of low-field MRI medical volumes ($\mathcal{D}$) are utilized simultaneously during training using the following multi-task loss function:

$$\mathcal{L}^i(\theta_i; \mathcal{X}_i) := \mathcal{L}_{\text{SEG}}^i(\theta_i; \mathcal{X}_i) + \lambda_m \mathcal{L}_{\text{M}}^i(\theta_i; \mathcal{X}_i) + \lambda_c \mathcal{L}_{\text{ADV}}^i(\theta_i; \mathcal{X}_i) \,, \tag{6}$$

where $\mathcal{L}_{\text{SEG}}^i$, $\mathcal{L}_{\text{M}}^i$ and $\mathcal{L}_{\text{ADV}}^i$ denote the Segmentation loss, the Masked Spatial Cross-Entropy loss, and the Adversarial loss, respectively. Here, λ_m and λ_c denote weighted parameters to control individual loss terms during the training. We set $\lambda_m = 0.3$ and $\lambda_c = 0.01$ in all our experiments.

The Segmentation Loss ($\mathcal{L}_{\text{SEG}}$). The Segmentation loss drives each segmentation network to produce prediction masks for labeled data that closely match the ground truth masks. We define the total Segmentation loss as the sum of the Cross-Entropy loss and Dice loss, both calculated voxel-wise. The primary segmentation loss is defined as follows:

$$\mathcal{L}_{\text{CE}}^i(\theta_i; \mathcal{X}_i) = \sum_{i \in m} \left[-\mathbb{E}_{(\mathbf{X},\mathbf{Y}) \sim \mathcal{X}_i} \left[\langle \mathbf{Y}, \log \left(\mathcal{F}_i(\mathbf{X}, i, \hat{\mathbf{Y}}_{i-1}, \mathbf{Z}_{i-1}) \right) \rangle \right] \right], \quad (7)$$

$$\mathcal{L}_{\text{DICE}}^i(\theta_i; \mathcal{X}_i) = \sum_{i \in m} \left[1 - \mathbb{E}_{(\mathbf{X},\mathbf{Y}) \sim \mathcal{X}_i} \left[\frac{2 \langle \mathbf{Y}, \mathcal{F}_i(\mathbf{X}, i, \hat{\mathbf{Y}}_{i-1}, \mathbf{Z}_{i-1}) \rangle}{\|\mathbf{Y}\|_1 + \|\mathcal{F}_i(\mathbf{X}, i, \hat{\mathbf{Y}}_{i-1}, \mathbf{Z}_{i-1})\|_1} \right] \right], \quad (8)$$

$$\mathcal{L}_{\text{SEG}}^i(\theta_i; \mathcal{X}_i) = \mathcal{L}_{\text{CE}}^i(\theta_i; \mathcal{X}_i) + \mathcal{L}_{\text{DICE}}^i(\theta_i; \mathcal{X}_i), \quad (9)$$

The Adversarial Loss ($\mathcal{L}_{\text{ADV}}$). In our training pipeline, we use a critic network which has the functionality of $\psi : [0,1]^{H \times W \times D} \to [0,1]^{H \times W \times D}$ that helps the segmentation network to generate realistic segmentation masks using min-max game as defined in Eq. 5. The adversarial loss for the training segmentation network is defined as:

$$\mathcal{L}_{\text{ADV}}^i(\theta_i; \mathcal{X}) := -\mathbb{E}_{(\mathbf{X},\mathbf{Y}) \sim \mathcal{X}_i} \left[\sum_{a \in H} \sum_{b \in W} \sum_{c \in D} \log \left(\psi(\mathcal{F}_i(\mathbf{X}, \hat{\mathbf{Y}}))[a,b,c] \right) \right], \quad (10)$$

The Masked Spatial Loss ($\mathcal{L}_{\text{M}}$). Further, we integrate a spatial masked CE loss to train the model via uncertainty, which leads to co-learn from each model's features. Here, we make the masked segmentation prediction map by binarizing the confidence map using a predefined threshold of $T = 0.2$. The masked loss is defined as follows:

$$\mathcal{L}_{\text{M}}^i(\theta_i; \mathcal{X}_i) := -\mathbb{E}_{(\mathbf{X},\mathbf{Y}) \sim \mathcal{X}_i} \Big[\sum_{a,b,c} \mathbb{1}(\psi(\mathcal{F}(\mathbf{X}, \hat{\mathbf{Y}})[a,b,c] > T)$$
$$\mathbf{Y}[a,b,c] \log \left(\mathcal{F}(\mathbf{X}, \hat{\mathbf{Y}})[a,b,c] \right) \Big], \quad (11)$$

The Critic Loss ($\mathcal{L}_{\text{C}}$). To train the critic network, we use segmentation masks and their ground truth masks. We define the adversarial loss as maximizing the log-likelihood as:

$$\mathcal{L}_C(\theta_c; \mathcal{D}) := \mathbb{E}_{(\mathbf{X},\mathbf{Y}) \sim \mathcal{X}_1} \Bigg[\sum_{a \in H} \sum_{b \in W} \sum_{c \in D} \Big\{ \eta \log \big(\psi(\mathbf{Y})[a,b,c] \big) + (1-\eta)$$
$$\log \big(1 - \psi_i(\mathcal{F}_i(\mathbf{X}, \hat{\mathbf{Y}}))[a,b,c] \big) \Big\} \Bigg] +$$
$$\mathbb{E}_{(\mathbf{X},\mathbf{Y}) \sim \mathcal{X}_2} \Bigg[\sum_{a \in H} \sum_{b \in W} \sum_{c \in D} \Big\{ \eta \log \big(\psi(\mathbf{Y})[a,b,c] \big)$$
$$+ (1-\eta) \log \big(1 - \psi_i(\mathcal{F}_i(\mathbf{X}, \hat{\mathbf{Y}}))[a,b,c] \big) \Big\} \Bigg]. \quad (12)$$

where $\eta = 0$ when the sample is a prediction mask from a segmentation network, and $\eta = 1$ when the sample is obtained from the ground truth label distribution.

4 Experiments

4.1 Implementation Details

The min-max scaling was performed to standardize all volumes, followed by clipping intensity values. Images were then cropped to a fixed patch size of $128 \times 128 \times 128$ by removing unnecessary background pixels [9–11]. The LoFiHippSeg model is implemented in PyTorch and trained using a single NVIDIA A100 GPU with 80GB of memory. For training the segmentation networks, we utilized the batch size of 4 and the SGD optimizer with a learning rate of 0.01 and a momentum of 0.9. The critic network was trained with the AdamW optimizer, which had a learning rate of 0.0001. We applied a cosine annealing scheduler to all networks throughout the training process. The training was conducted alternately between the segmentation networks and the critic. The critic is not used during inference, thereby avoiding additional computational overhead. We split the training dataset into a training set (76%) of 60 MR volumes for training and a validation set (24%) of 19 MR volumes for validation. The best-performing model for the validation set is saved as the best model for official validation and testing phase evaluation. The LISA 2024 validation dataset contains 12 MR volumes, and the Synapse portal conducts the evaluation. In the inference phase, the original volume was re-scaled using min-max normalization scaling and fed forward through the LoFiHippSeg model. The LoFiHippSeg model uses ensembled prediction from dual views as the final prediction during inference.

4.2 Evaluation Metrics

We will utilize five metrics for evaluating hippocampi segmentation predictions: Dice Similarity Coefficient (DSC), Hausdorff Distance (HD), HD95, Average Symmetric Surface Distance (ASSD), and Relative Volume Error (RVE). These metrics will be computed separately for the left and right hippocampus, and the results will be averaged for each patient case [6].

Table 1. Validation Phase Quantitative Comparison with VNet.

Metric	VNet [7]			LoFiHippSeg		
	Left	Right	Average	Left	Right	Average
DSC	0.69±0.23	0.72±0.15	0.70±0.19	**0.70±0.23**	**0.74±0.15**	**0.72±0.19**
HD	10.56±16.78	3.65±1.02	7.10±8.32	**5.99±9.12**	**3.49±1.18**	**4.74±4.62**
HD95	**2.11±1.95**	1.88±0.78	**1.99±1.33**	2.16±2.03	**1.86±0.92**	2.01±1.45
ASSD	0.87±1.19	0.61±0.39	0.74±0.78	**0.87±1.33**	**0.59±0.44**	**0.73±0.88**
RVE	0.18±0.10	**0.14±0.10**	0.16±0.07	**0.14±0.12**	0.14±0.12	**0.14±0.10**

Table 2. Comparison of segmentations by VNet and LoFiHippSeg across axial, sagittal, and coronal views during Validation phase.

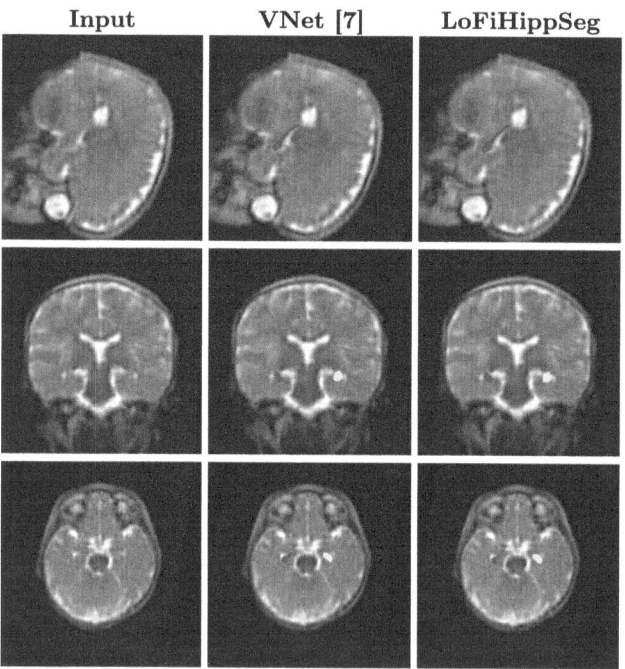

Table 3. Validation Phase Quantitative Analysis on Dual-Views of LoFiHippSeg.

Metric	LoFiHippSeg View 1			LoFiHippSeg View 2		
	Left	Right	Average	Left	Right	Average
DSC	0.69±0.23	0.73±0.16	0.71±0.19	0.69±0.24	0.73±0.13	0.71±0.18
HD	8.21±15.89	3.52±0.98	5.86±7.92	5.80±8.03	6.61±11.06	6.20±6.43
HD95	2.21±2.20	1.88±0.89	2.04±1.52	2.29±2.02	1.92±1.04	2.10±1.51
ASSD	0.94±1.43	0.59±0.45	0.77±0.94	0.89±1.28	0.63±0.40	0.76±0.83
RVE	0.19±0.12	0.15±0.12	0.17±0.08	0.15±0.09	0.12±0.11	0.14±0.08

4.3 Experimental Results

We evaluated the method's performance using the LISA 2024 Validation Phase evaluation portal, and results are shown in Table 1. From the results, it can be seen that the proposed dual-view setting helps in better feature retrieval over a single VNet segmentation model (See Table 3). Qualitative comparison of generated prediction masks are illustrated in the Table 2.

Comparison with VNet. The performance comparison between LoFiHippSeg and VNet [7], shown in Table 1, indicates a modest improvement in segmentation accuracy for the LoFiHippSeg model. Specifically, LoFiHippSeg achieved a slightly higher DSC (0.72±0.19) compared to VNet (0.70±0.19), demonstrating an enhanced ability to correctly classify hippocampal regions, particularly for the right hippocampus (0.74±0.15 versus 0.72±0.15 in VNet). This increase, though marginal, signifies that the LoFiHippSeg model can better delineate hippocampal structures, possibly due to its enhanced learning capabilities from low-field MRI scans. In terms of boundary accuracy, LoFiHippSeg also outperformed VNet in HD (4.74±4.62 versus 7.10±8.32 for VNet). The reduction in HD suggests that the segmentation boundaries produced by LoFiHippSeg are more precise, particularly for the left hippocampus, where the HD decreased from 10.56±16.78 in VNet to 5.99±9.12 in LoFiHippSeg. This reduction could imply fewer outliers in the boundary predictions by the LoFiHippSeg model. However, HD95 showed relatively comparable values between the two models, indicating that extreme outliers in the segmentation were not substantially different. Regarding ASSD, which measures the average surface distance between the predicted and true segmentations, both models performed similarly, with the overall averages almost identical (0.73±0.88 for LoFiHippSeg and 0.74±0.78 for VNet). This metric aligns with the HD observations, indicating that while the general boundary accuracy has improved, there is room for further refinement. The RVE, a volumetric measure, shows slight improvement for LoFiHippSeg (0.14±0.10) compared to VNet (0.16±0.07). This indicates that LoFiHippSeg produces more accurate volume estimations, which is critical for clinical applications where hippocampal volume is a biomarker for neurodegenerative conditions. As illustrated in Fig. 2, the qualitative differences in segmentations are not immediately noticeable to the human eye. However, it is evident that LoFiHippSeg accurately captured some misclassified regions compared to the single VNet.

Dual-View Architecture Analysis. Table 3 provides further insights into the performance of the dual-view architecture of LoFiHippSeg. Both views exhibit similar performance across most metrics, indicating robustness in the model's segmentation ability regardless of the view utilized. DSC values for both views are nearly identical (0.71±0.19 for View 1 and 0.71±0.18 for View 2), which highlights the stability of the model's segmentation performance from different perspectives. One key observation is the difference in HD between the two views.

View 1 exhibits a lower HD (5.86±7.92) compared to View 2 (6.20±6.43). While this difference is not substantial, it suggests that the first view may provide slightly more precise boundary delineation, particularly for the left hippocampus, which shows a notable decrease in HD for View 1 (8.21±15.89 versus 5.80±8.03 for View 2). The HD95 and ASSD metrics, however, remain consistent across both views, reinforcing the robustness of the segmentation performance. Interestingly, the RVE metric shows a marginal improvement in View 2 (0.14±0.08) compared to View 1 (0.17±0.08). This could indicate that the second view is more effective in achieving accurate volumetric estimations, particularly for the right hippocampus (0.12±0.11 for View 2 versus 0.15±0.12 for View 1). These complementary strengths of each view suggest that a combined approach leveraging both views could potentially yield even better performance.

4.4 Ablation Study

One of the central claims of the proposed model is its use of high-frequency masking to generate high-frequency images of low-field images. In Fig. 3, we illustrate how varying cutoff values produce different high-pass filters and the corresponding feature difference maps between the low-field and high-frequency images. These qualitative visualizations show that the feature difference decreases as the cutoff value increases. A more noticeable feature difference emerges at lower cutoff values, even though it may not be easily perceived by the naked eye, as demonstrated with a cutoff of 0.05. However, fine details are not as clearly visible

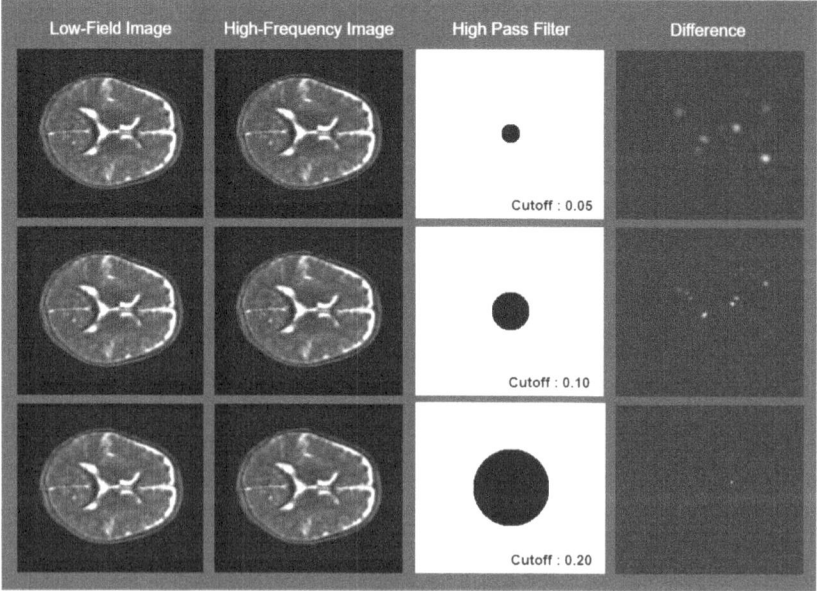

Fig. 3. Feature difference between Low-field Image and High-frequency Image.

Table 4. Ablation Study of cutoff value.

Metric	Cutoff=0.05	Cutoff=0.10	Cutoff=0.20
Average DSC	0.71±0.18	0.72±0.19	0.70±0.19
Average HD	9.12±12.58	4.74±4.62	5.90±7.93
Average HD95	2.04±1.54	2.01±1.45	2.16±1.59
Average ASSD	0.75±0.85	0.73±0.88	0.77±0.91
Average RVE	0.14±0.10	0.14±0.10	0.20±0.10

at this value as they are with a cutoff of 0.1. In our ablation study, we evaluated the model's performance using these three cutoff values, with the results summarized in Table 4. The findings indicate that a cutoff value of 0.1 yields the best performance compared to the other two.

4.5 Discussion

The proposed LoFiHippSeg outperforms a single VNet trained on low-field MRI images. While the model achieves better results, it does present certain limitations, such as increased computational complexity. However, with the ongoing technological advancements, computational complexity is becoming less of a constraint. We believe that incorporating more advanced segmentation models over VNet could further enhance segmentation performance.

5 Conclusion

In this study, we introduced a novel deep-learning approach for the automatic segmentation of bilateral hippocampi in low-field MRIs, addressing a critical need in diagnosing and studying cognitive and memory functions in neurodevelopmental disorders. By adapting recent advancements in infant brain segmentation to low-field MRIs, our method extends the accessibility of essential diagnostic tools to underserved communities, promoting equitable healthcare for all children.

Acknowledgments. This study was funded by the Australian Research Council Discovery Program DP210101863.

Disclosure of Interests. The authors have no competing interests.

References

1. Blum, A., Mitchell, T.: Combining labeled and unlabeled data with co-training. In: Proceedings of the Eleventh Annual Conference on Computational Learning Theory, pp. 92–100 (1998)

2. Çiçek, Ö., Abdulkadir, A., Lienkamp, S.S., Brox, T., Ronneberger, O.: 3D u-net: learning dense volumetric segmentation from sparse annotation. In: International Conference on Medical Image Computing and Computer-assisted Intervention, pp. 424–432. Springer (2016)
3. Dasgupta, S., Littman, M.L., McAllester, D.: PAC generalization bounds for co-training. In: Advances in Neural Information Processing Systems, pp. 375–382 (2002)
4. Deoni, S.C., et al.: Development of a mobile low-field MRI scanner. Sci. Rep. **12**(1), 5690 (2022)
5. Liang, Z., Guo, X., Noble, J.A., Kamnitsas, K.: Itermask2: iterative unsupervised anomaly segmentation via spatial and frequency masking for brain lesions in MRI. arXiv preprint arXiv:2406.02422 (2024)
6. Maier-Hein, L., et al.: Bias: transparent reporting of biomedical image analysis challenges. Med. Image Anal. **66**, 101796 (2020)
7. Milletari, F., Navab, N., Ahmadi, S.A.: V-net: fully convolutional neural networks for volumetric medical image segmentation. In: 3DV 2016, pp. 565–571 (2016)
8. Peiris, H., Chen, Z., Egan, G., Harandi, M.: Duo-segnet: adversarial dual-views for semi-supervised medical image segmentation. In: International Conference on Medical Image Computing and Computer-Assisted Intervention, pp. 428–438. Springer (2021)
9. Peiris, H., Chen, Z., Egan, G., Harandi, M.: Reciprocal adversarial learning for brain tumor segmentation: a solution to brats challenge 2021 segmentation task. In: Crimi, A., Bakas, S. (eds.) Brainlesion: Glioma, Multiple Sclerosis, Stroke and Traumatic Brain Injuries, pp. 171–181. Springer International Publishing, Cham (2022)
10. Peiris, H., Hayat, M., Chen, Z., Egan, G., Harandi, M.: Hybrid window attention based transformer architecture for brain tumor segmentation. In: International MICCAI Brainlesion Workshop, pp. 173–182. Springer (2022)
11. Peiris, H., Hayat, M., Chen, Z., Egan, G., Harandi, M.: A robust volumetric transformer for accurate 3D tumor segmentation. In: International Conference on Medical Image Computing and Computer-Assisted Intervention, pp. 162–172. Springer (2022)
12. Peiris, H., Hayat, M., Chen, Z., Egan, G., Harandi, M.: Uncertainty-guided dual-views for semi-supervised volumetric medical image segmentation. Nat. Mach. Intell. **5**, 724–738 (2023)
13. Sheth, K.N., et al.: Assessment of brain injury using portable, low-field magnetic resonance imaging at the bedside of critically ill patients. JAMA Neurol. **78**(1), 41–47 (2021)
14. Smith, S.M., et al.: Advances in functional and structural MR image analysis and implementation as FSL. Neuroimage **23**, S208–S219 (2004)
15. Xu, C., Tao, D., Xu, C.: A survey on multi-view learning. arXiv preprint arXiv:1304.5634 (2013)

Open Access This chapter is licensed under the terms of the Creative Commons Attribution 4.0 International License (http://creativecommons.org/licenses/by/4.0/), which permits use, sharing, adaptation, distribution and reproduction in any medium or format, as long as you give appropriate credit to the original author(s) and the source, provide a link to the Creative Commons license and indicate if changes were made.

The images or other third party material in this chapter are included in the chapter's Creative Commons license, unless indicated otherwise in a credit line to the material. If material is not included in the chapter's Creative Commons license and your intended use is not permitted by statutory regulation or exceeds the permitted use, you will need to obtain permission directly from the copyright holder.

Infant Hippocampal Segmentation in Ultra-Low-Field MRI Using External Datasets with Diverse Field Strengths

Weichen Zhou[1], Jingyu Li[1], Xi Wang[2,3], Yuwan Wang[1], and Mengye Lyu[1(✉)]

[1] Shenzhen Technology University, Shenzhen, China
lvmengye@sztu.edu.cn
[2] The Chinese University of Hong Kong, Hong kong, China
[3] Zhejiang Lab, Hangzhou, China

Abstract. Ultra-low-field (uLF) MRI is a promising technology for expanding access to MRI in resource-limited settings, but it presents significant image analysis challenges, particularly in the segmentation of small brain structures. One such structure is the hippocampus, which is crucial for monitoring the neurodevelopment after birth. In this study, we developed a hippocampal segmentation method using the nnUNet framework, exploring various training strategies, including using additional labeled high-field (HF) and unlabeled low-field (LF) data. Our results show that integrating external datasets improves segmentation accuracy over using uLF data alone, even with substantial differences in imaging parameters and field strengths. This approach highlights the importance of leveraging diverse datasets to enhance the performance of segmentation models in low-quality imaging modalities, potentially leading to better diagnostic capabilities in challenging clinical environments.

Keywords: Neonatal Brain · Pediatric Imaging · Hippocampus

1 Introduction

Magnetic Resonance Imaging (MRI) has revolutionized the field of medical imaging, providing a non-invasive, non-ionizing, and highly detailed visualization of the human body. Its ability to produce high-resolution, multi-contrast images has made it indispensable for diagnosing and monitoring a wide range of conditions, particularly in the brain. However, despite its benefits, traditional high-field MRI systems (1.5T and 3T) remain inaccessible in many low and middle-resource countries due to their high costs, complex infrastructure requirements, and the need for specialized settings [11,18]. This limited accessibility is a significant barrier to equitable healthcare, as it restricts the availability of critical diagnostic tools to resource-rich settings. To address this disparity, low-field (LF) MRI systems, especially Ultra-Low-Field (uLF) MRI systems below 0.1T, have been

W. Zhou and J. Li—Equal contribution.

developed [4,5,12,17,22]. These systems offer a more affordable, portable, and infrastructure-light alternative to conventional MRI, making them particularly suitable for deployment in resource-constrained environments.

Despite the advantages of uLF MRI, these systems face significant challenges, especially in terms of image quality. The lower magnetic field strength inherent in uLF MRI results in a decreased signal-to-noise ratio (SNR), leading to images with lower resolution and contrast compared to those produced by high-field MRI systems [1,16]. This presents particular difficulties for image analysis tasks that require precise anatomical delineation, such as the segmentation of small and complex structures within the brain.

One such structure is the hippocampus, a critical component of the brain involved in memory formation, spatial navigation, and emotional regulation [7]. In neonates and infants, the hippocampus plays a pivotal role in cognitive development. Research has demonstrated that the volume and integrity of the hippocampus in early life can significantly influence cognitive outcomes [2,19,20], underscoring the importance of accurately assessing and monitoring the hippocampus during early childhood.

However, accurately segmenting the hippocampus in MRI images is a challenging task [6], even more so in the context of uLF MRI [3,9]. The hippocampus is small, varies in shape, and has low contrast with surrounding tissues, making it difficult to delineate using conventional segmentation techniques. Moreover, pediatric brain MRI typically exhibits lower tissue contrast, which can vary significantly over the first few years of life, posing additional challenges for segmentation tasks. Traditional methods, which often rely on manual feature extraction and expert annotation, are time-consuming, prone to error, and do not scale well. The advent of deep learning has provided new opportunities to automate the segmentation process, with models trained on large datasets showing promising results in accurately delineating complex anatomical structures. Among these models, nnUNet [10] has emerged as an advanced tool in the field of medical image segmentation. Known for its robustness, adaptability, and superior performance across a range of medical imaging tasks, nnUNet offers a well-optimized, highly customizable pipeline that has won numerous medical image segmentation challenges. This makes it an ideal foundation for developing a hippocampal segmentation method tailored to the unique challenges posed by uLF MRI.

While the choice of segmentation algorithm is critical, the quality and diversity of the training data are equally, if not more, important. In the case of uLF MRI, where image quality is inherently lower, the ability of a model to generalize across different datasets becomes paramount. High-quality training data that captures a wide range of anatomical variations, imaging conditions, and contrast levels may effectively enhance the performance of segmentation models. This is especially true when the imaging modality is less conventional, as in uLF MRI.

In this study, we aim to develop an automatic method for segmenting the bilateral hippocampi from 0.064T MRI [4,9,17] images using the nnUNet framework. Our approach focuses on leveraging external datasets to improve segmentation performance. The planned experiments include several strategies using multiple datasets. By improving the reliability and accuracy of hippocampal segmentation, our work could contribute to better diagnostic capabilities in resource-constrained settings, where access to high-field MRI is limited.

2 Methods

2.1 Model Architecture

We employed the nnUNet [10] framework for hippocampal segmentation. nnUNet is a state-of-the-art, self-configuring deep learning framework designed for medical image segmentation. It automatically adapts its architecture and training pipeline based on the input data characteristics, providing a robust baseline for a wide range of segmentation tasks. nnUNet's architecture is based on the U-Net structure, featuring an encoder-decoder design with skip connections, which is well-suited for segmenting small, complex anatomical structures like the hippocampus.

2.2 Datasets

Our study primarily focuses on the dataset provided by the Low-field Pediatric Brain Magnetic Resonance Image Segmentation and Quality Assurance (LISA) Challenge, consisting of 0.064T uLF MRI images. Additionally, we incorporated high-field MRI data from the 2024 China National Biomedical Engineering Innovation Design Competition for College Students competition (BME24) and 0.3T MRI dataset M4Raw [13] to enhance model training. The LISA and BME24 datasets included labeled hippocampal regions to facilitate supervised learning, whereas the M4Raw dataset contains only low-field MRI images without any segmentation labels. Typical images of the three datasets are presented in Fig. 1. Below are detailed description for the three datasets.

The LISA dataset, as provided by the LISA organizers, was acquired from infants using the Hyperfine SWOOP 0.064T MRI scanner, which is designed for portable brain imaging in pediatric settings. The uLF images were captured with a spin echo sequence (TR 1.5 s, TE 5ms, TI 400ms) and were subsequently registered to corresponding high-field MRI images. The preprocessing involved registering the uLF images to a standardized high-field T2 template using ANTs multivariate template construction and FLIRT from the FSL toolbox, ensuring consistent anatomical alignment and facilitating accurate hippocampal segmentation. The dataset includes 79 training and 12 validation files (samples), with bilateral hippocampi segmentation labels reviewed by an expert medical image evaluator. Besides the official validation set of 12 samples, whose segmentation labels were not released, we randomly split the training set to obtain 13 samples as the internal validation set, remaining 66 samples were used for model training.

The BME24 dataset is part of the 2024 China National Biomedical Engineering Innovation Design Competition for College Students, organized by the

Chinese Society of Biomedical Engineering. The dataset comprises high-field T1-weighted Magnetization Prepared Rapid Gradient Echo Imaging (MPRAGE) images, with acquisition matrix [224 × 256 × 176] from 58 adult subjects, provided with bilateral hippocampal segmentation labels. The dataset is openly accessible and can be downloaded from https://pan.baidu.com/s/1x29I5swjtRy-C_R96EUnzw?pwd=yyna. Further details about the competition are available on the official website https://www.bmedesign.cn/. It should be noted that the BME24 dataset only includes T1-weighted images and does not contain T2-weighted data.

The M4Raw dataset [13] is an open-access dataset created to support the development of advanced methodologies in low-field MRI. The dataset comprises multi-channel brain k-space data from 183 healthy adult volunteers, acquired using a 0.3 T whole-body MRI system. It originally includes T1-weighted, T2-weighted, and FLAIR images, each with an in-plane resolution of approximately 1.2 mm and a through-plane resolution of 5 mm. For this study, we specifically utilized the T2-weighted data (magnitude images) to maintain consistency with the T2-weighted data from the LISA dataset. The lack of segmentation labels in the M4Raw dataset renders it unsuitable for supervised training. However, it can still provide information about low-field MRI data distribution, which can be leveraged through self-supervised or unsupervised learning techniques.

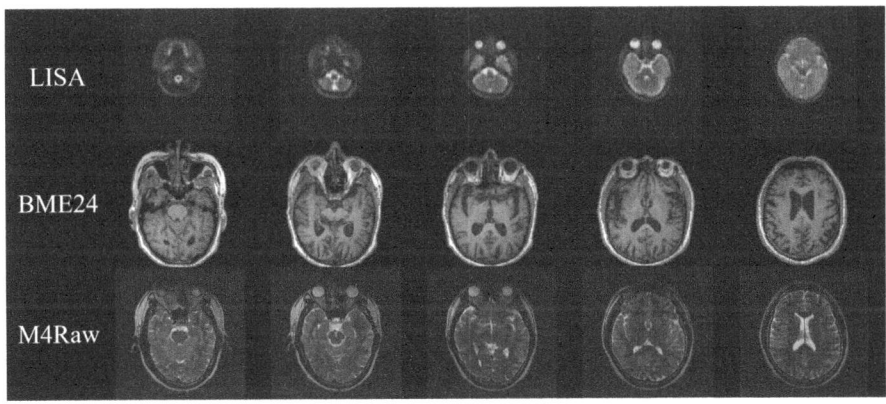

Fig. 1. Typical images of the three datasets utilized in this study.

2.3 Training Strategies

We conducted a series of experiments to evaluate the impact of different training strategies on segmentation performance:

1. **uLF-Only Training**: The baseline model was trained solely on the LISA dataset, containing only 0.064T uLF MRI images.

2. **Mixed Training**: In this experiment, we combined the LISA dataset with the BME24 dataset, which included T1-weighted high-field MRI images. The model was trained on this mixed dataset to investigate whether incorporating higher-quality images could improve segmentation accuracy in uLF MRI.
3. **HF Transfer**: We pre-trained the model using the BME24 high-field dataset and then fine-tuned it on the LISA uLF data. This approach aimed to leverage the detailed structural information available in high-field images before adapting the model to the lower-quality uLF data.
4. **Mixed Transfer**: We first pre-trained the model on a mixed dataset (LISA + BME24), followed by a second fine-tuning phase using only the LISA data. This sequential approach was designed to refine the model's performance on uLF images by focusing specifically on their unique characteristics after broad initial training.
5. **Pseudo-labeling**: The best-performing model from the previous experiments was used to generate pseudo-labels on an additional set of LF MRI images (M4Raw T2-weighted images). These pseudo-labeled images were then added to the training data to further enhance the model's segmentation capabilities.

2.4 Evaluation Metrics

We employ four metrics [15,21], Dice Similarity Coefficient (DSC), Hausdorff Distance (HD), Average Symmetric Surface Distance (ASSD), and Relative Volume Error (RVE) on the hippocampi segmentation predictions. These metrics will be calculated for the left and right hippocampus, separately, but then averaged along a patient case. The calculation of these metrics were performed on the official platform based on submitted segmentation files.

2.5 Implementation Details

The experiments were implemented using Python and the PyTorch deep learning library. Training was conducted on a GPU Server. The training protocols are presented in Table 1.

Table 1. Training protocols

Network initialization	"he" normal initialization
Batch size	2
Patch size	112×160×128
Total epochs	50 or 200
Optimizer	SGD with nesterov momentum ($\mu = 0.99$)
Loss function	cross-entropy loss & dice loss
Initial learning rate (lr)	0.01
Lr decay schedule	halved by 200 epochs
Training time	1 h

The hardware configuration and development environments are presented in Table 2.

Table 2. Development environments and requirements.

System	Ubuntu 22.04
CPU	AMD EPYC 7453 28-Core Processor
RAM	24×32GB
GPU (number and type)	Four NVIDIA GeForce RTX 4090 24G
CUDA version	12.1
Programming language	Python 3.10.10
Deep learning framework	torch 2.1.2, torchvision 0.16.2

3 Results

The quantitative results across different training strategies are summarized in Tables 3, 4, and 5, for average, left, and right hippocampal segmentation, respectively.

As revealed by the average results in Table 3, the baseline model, trained solely on the uLF data (uLF-Only Training), achieved a DSC of 0.66 ± 0.22. Incorporating high-field MRI data through Mixed Training resulted in a slight improvement, with a DSC of 0.68 ± 0.19. Transfer learning strategies-HF Transfer and Mixed Transfer-further improved segmentation accuracy, with DSCs of 0.70 ± 0.17 and 0.71 ± 0.18, respectively. These strategies also reduced HD and ASSD, indicating better boundary accuracy. However, the Pseudo-Labeling approach showed a decrease in performance, with a DSC of 0.64 ± 0.27.

The separate results for the left and right hippocampi are largely consistent with the averaged results. However, Mixed Training appears to be particularly effective on the right hippocampus, achieving a leading DSC of 0.74 ± 0.14.

Table 3. Metrics evaluated on official validation set (bilateral average)

Method	DSC	HD	HD95	ASSD	RVE
uLF-Only Training	0.66 ± 0.22	9.95 ± 11.96	6.16 ± 9.80	1.44 ± 2.09	0.19 ± 0.13
Mixed Training	0.68 ± 0.19	7.87 ± 10.54	4.15 ± 7.21	1.16 ± 1.54	0.19 ± 0.16
HF Transfer	0.70 ± 0.17	8.15 ± 9.60	3.73 ± 5.41	0.80 ± 0.64	0.16 ± 0.06
Mixed Transfer	0.71 ± 0.18	9.66 ± 10.64	3.99 ± 6.52	0.81 ± 0.76	0.13 ± 0.07
Pseudo-Labeling	0.64 ± 0.27	10.11 ± 14.20	5.93 ± 12.51	3.99 ± 11.14	0.16 ± 0.08

The visualization of typical segmentation results from our internal validation set is presented in Fig. 2, as it has known segmentation labels (ground truth).

Table 4. Metrics evaluated on official validation set (left hippocampus)

Method	DSC	HD	HD95	ASSD	RVE
uLF-Only Training	0.64 ± 0.25	14.96 ± 20.72	8.90 ± 15.34	1.99 ± 2.92	0.17 ± 0.12
Mixed Training	0.62 ± 0.25	9.23 ± 15.81	3.23 ± 2.95	1.09 ± 1.26	0.23 ± 0.21
HF Transfer	0.67 ± 0.24	9.41 ± 16.13	2.70 ± 1.94	0.90 ± 1.03	0.19 ± 0.09
Mixed Transfer	0.68 ± 0.23	15.39 ± 20.92	6.00 ± 12.80	1.02 ± 1.20	0.15 ± 0.11
Pseudo-Labeling	0.60 ± 0.29	12.56 ± 19.64	6.23 ± 12.44	4.18 ± 11.08	0.19 ± 0.12

Table 5. Metrics evaluated on official validation set (right hippocampus)

Method	DSC	HD	HD95	ASSD	RVE
uLF-Only Training	0.68 ± 0.23	4.93 ± 6.00	3.42 ± 5.77	0.89 ± 1.30	0.22 ± 0.23
Mixed Training	0.74 ± 0.14	6.51 ± 12.20	5.06 ± 12.09	1.23 ± 2.53	0.15 ± 0.14
HF Transfer	0.73 ± 0.12	6.89 ± 10.83	4.75 ± 9.89	0.69 ± 0.46	0.12 ± 0.08
Mixed Transfer	0.73 ± 0.14	3.93 ± 0.92	1.98 ± 0.68	0.59 ± 0.35	0.11 ± 0.08
Pseudo-Labeling	0.68 ± 0.25	7.66 ± 12.48	5.62 ± 12.62	3.80 ± 11.23	0.13 ± 0.10

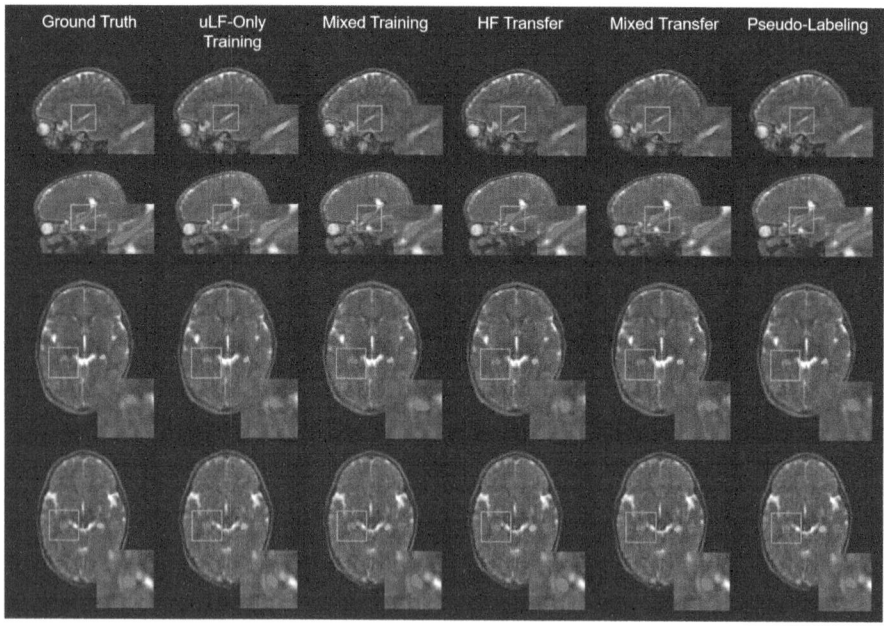

Fig. 2. Visualization of the horizontal section and sagittal section segmentation results.

4 Discussion

Our study explored various training strategies for hippocampal segmentation using a combination of targeted uLF data and data acquired at other MRI scanners with diverse field strengths and site.

We observed that mixed training using both T2 uLF images and T1 HF images resulted in slightly better segmentation performance compared to training on uLF images alone. This improvement, though modest, suggests that the inclusion of HF data provides additional structural information that can enhance the model's ability to generalize to uLF images. The structural details captured in the HF T1-weighted images, despite being of a different contrast type, might help the model learn more robust features, which can be beneficial when segmenting the hippocampus in lower-quality uLF images.

Both transfer learning strategies-HF Transfer and Mixed Transfer-showed improved performance compared to uLF-Only Training. HF Transfer, which involved pre-training on the BME24 high-field data followed by fine-tuning on the LISA uLF data, demonstrated that the model could effectively transfer learned features from the HF domain to improve segmentation in the uLF domain. Mixed Transfer, which included initial mixed training followed by fine-tuning on uLF data, further refined the model's performance by focusing on the unique characteristics of uLF images. These results indicate that transfer learning, particularly when leveraging detailed structural information from HF images, can significantly enhance the model's adaptability and accuracy in segmenting uLF MRI images.

Despite the expectation that the Pseudo-Labeling strategy using the M4Raw T2-weighted data would enhance segmentation accuracy, as shown effective by some previous studies [8,14], it did not lead to direct improvements in this study. A possible explanation for our outcome is the potential inaccuracy of the pseudo-labels generated by the model, which may have introduced noise rather than enhancing the training process. The lack of true segmentation labels in the M4Raw dataset likely limited the effectiveness of this approach, highlighting the importance of label quality in supervised learning tasks.

In conclusion, our findings underscore the value of incorporating external MRI data through mixed training and transfer learning to improve segmentation performance in low-field MRI. However, the effectiveness of pseudo-labeling needs future investigation.

Acknowledgments. The presented solution by Synapse Team wanziya (ID 3505152) is for submission to LISA 2024. This work was in part supported by the National Natural Science Foundation of China (No. 62101348), Shenzhen Science and Technology Program (Shenzhen Higher Education Stable Support Program, No. 20220716111838002), and Natural Science Foundation of Top Talent of Shenzhen Technology University (No. GDRC202134).

Disclosure of Interests. The authors have no competing interests.

References

1. Arnold, T.C., Freeman, C.W., Litt, B., Stein, J.M.: Low-field MRI: clinical promise and challenges. J. Magn. Reson. Imaging **57**(1), 25–44 (2023)
2. Beauchamp, M.H., et al.: Preterm infant hippocampal volumes correlate with later working memory deficits. Brain **131**(11), 2986–2994 (2008)
3. de Leeuw den Bouter, M., Ippolito, G., O'Reilly, T., Remis, R., van Gijzen, M., Webb, A.: Deep learning-based single image super-resolution for low-field MR brain images. Sci. Rep. **12**(1), 6362 (2022)
4. Deoni, S.C., et al.: Accessible pediatric neuroimaging using a low field strength MRI scanner. Neuroimage **238**, 118273 (2021)
5. Deoni, S.C., et al.: Development of a mobile low-field MRI scanner. Sci. Rep. **12**(1), 5690 (2022)
6. Dill, V., Franco, A.R., Pinho, M.S.: Automated methods for hippocampus segmentation: the evolution and a review of the state of the art. Neuroinformatics **13**, 133–150 (2015)
7. Eichenbaum, H., Otto, T., Cohen, N.J.: The hippocampus-what does it do? Behav. Neural Biol. **57**(1), 2–36 (1992)
8. Huang, S., Yang, H., Mei, L., Zhang, T., Liu, S., Lyu, M.: From whole-body to abdomen: streamlined segmentation of organs and tumors via semi-supervised learning and efficient coarse-to-fine inference. Lect. Notes Comput. Sci. **14544**, 22 (2024)
9. Iglesias, J.E., et al.: Quantitative brain morphometry of portable low-field-strength MRI using super-resolution machine learning. Radiology **306**(3), e220522 (2022)
10. Isensee, F., Jaeger, P.F., Kohl, S.A., Petersen, J., Maier-Hein, K.H.: nnU-Net: a self-configuring method for deep learning-based biomedical image segmentation. Nat. Methods **18**(2), 203–211 (2021)
11. Jalloul, M., et al.: MRI scarcity in low-and middle-income countries. NMR Biomed. **36**(12), e5022 (2023)
12. Liu, Y., et al.: A low-cost and shielding-free ultra-low-field brain MRI scanner. Nat. Commun. **12**(1), 7238 (2021)
13. Lyu, M., et al.: M4raw: a multi-contrast, multi-repetition, multi-channel MRI k-space dataset for low-field MRI research. Sci. Data **10**(1), 264 (2023)
14. Ma, J., et al.: Unleashing the strengths of unlabeled data in pan-cancer abdominal organ quantification: the flare22 challenge. arXiv:2308.05862 (2023)
15. Maier-Hein, L., et al.: Bias: transparent reporting of biomedical image analysis challenges. Med. Image Anal. **66**, 101796 (2020)
16. Marques, J.P., Simonis, F.F., Webb, A.G.: Low-field MRI: an MR physics perspective. J. Magn. Reson. Imaging **49**(6), 1528–1542 (2019)
17. Mazurek, M.H., et al.: Portable, bedside, low-field magnetic resonance imaging for evaluation of intracerebral hemorrhage. Nat. Commun. **12**(1), 5119 (2021)
18. Murali, S., et al.: Bringing MRI to low-and middle-income countries: directions, challenges and potential solutions. NMR Biomed. **37**, e4992 (2023)

19. Overfeld, J., et al.: Neonatal hippocampal volume moderates the effects of early postnatal enrichment on cognitive development. Dev. Cogn. Neurosci. **45**, 100820 (2020)
20. Thompson, D.K., et al.: Neonate hippocampal volumes: prematurity, perinatal predictors, and 2-year outcome. Ann. Neurol. **63**(5), 642–651 (2008)
21. Yeghiazaryan, V., Voiculescu, I.: Family of boundary overlap metrics for the evaluation of medical image segmentation. J. Med. Imaging **5**(1), 015006–015006 (2018)
22. Zhao, Y., et al.: Whole-body magnetic resonance imaging at 0.05 tesla. Science **384**(6696), eadm7168 (2024)

Open Access This chapter is licensed under the terms of the Creative Commons Attribution 4.0 International License (http://creativecommons.org/licenses/by/4.0/), which permits use, sharing, adaptation, distribution and reproduction in any medium or format, as long as you give appropriate credit to the original author(s) and the source, provide a link to the Creative Commons license and indicate if changes were made.

The images or other third party material in this chapter are included in the chapter's Creative Commons license, unless indicated otherwise in a credit line to the material. If material is not included in the chapter's Creative Commons license and your intended use is not permitted by statutory regulation or exceeds the permitted use, you will need to obtain permission directly from the copyright holder.

Task 1 and Task 2 Combined

Automated Quality Assessment Using Appearance-Based Simulations and Hippocampus Segmentation on Low-Field Paediatric Brain MR Images

Vaanathi Sundaresan[1](✉) and Nicola K Dinsdale[2]

[1] Department of Computational and Data Sciences, Indian Institute of Science, Bangalore 560012, India
vaanathi@iisc.ac.in

[2] Oxford Machine Learning in NeuroImaging Lab, Department of Computer Science, University of Oxford, Oxford, UK

Abstract. Understanding the structural growth of paediatric brains is a key step in the identification of various neuro-developmental disorders. However, our knowledge is limited by many factors, including the lack of automated image analysis tools, especially in Low and Middle Income Countries from the lack of high field MR images available. Low-field systems are being increasingly explored in these countries, and, therefore, there is a need to develop automated image analysis tools for these images. In this work, as a preliminary step, we consider two tasks: 1) automated quality assurance and 2) hippocampal segmentation, where we compare multiple approaches. For the automated quality assurance task a DenseNet combined with appearance-based transformations for synthesising artefacts produced the best performance, with a weighted accuracy of 82.3%, thus ranking in the 1st place in the LISA2024 Challenge. For the segmentation task, registration of an average atlas performed the best, with a final Dice score of 0.61. Our results show that although the images can provide understanding of large scale pathologies and gross scale anatomical development, there still remain barriers for their use for more granular analyses.

Keywords: Low-field MRI · Quality Assurance · Hippocampal Segmentation · deep learning

1 Introduction

Understanding the healthy development of structures within the paediatric brain is vital for the identification of neuro-developmental disorders. So far, application of image analysis tools on large scale MR imaging studies (e.g., the UK Biobank [17]) for automated analysis of healthy structural development of the adult brain has been well described. However, the majority of existing image analysis tools struggle with paediatric brains due to the poor gray matter/white matter differentiation and rapid growth and change observed in brain anatomical structures. This, combined with the challenges of imaging children, such as the need to minimise the movement of infants, critically limits our ability to understand the structural development of the paediatric brain.

This knowledge gap is further increased when considering populations from Low and Middle Income Countries (LMICs) where high field MRI systems (1.5T / 3T) are rare due to the costs involved. Therefore, few MR imaging studies have considered LMIC populations. To fill this gap, Hyperfine SWOOP scanners are being tested - although at 0.064T the scanners offer much lower image quality, low-field MRI offers portability, cost-effectiveness and removes the need to sedate children during the scan. This improves the potential to increase the availability of MR imaging in underrepresented research communities [13].

Existing image analysis tools (e.g., FSL, Freesurfer) have been primarily developed for 1.5 and 3T MR images of adults: the large domain shift between these images and the low-field paediatric images we are considering mean that the existing tools are unlikely to give high quality results [6]. Therefore, there is a need to develop domain specific tools, designed specifically for the low-field images and paediatric populations. Deep learning (DL) based methods are generally good candidates for producing analysis tools, such as segmentation methods, for low-field paediatric images, due to their ability to identify underlying discriminative patterns in images and create both local- and global-level associations between voxel values [12]. Accurate segmentations of subcortical brain structures are essential for volumetric and morphological assessment and the monitoring of healthy brain development, but manual segmentation is time-consuming and error-prone [3]. The hippocampus is a grey matter structure located within the medial temporal lobe memory circuit, and is associated with numerous conditions including depression, Alzheimer's disease, psychosis and epilepsy [11]. Therefore, accurate delineation of the hippocampus is essential to enable volumetric and morphological assessment for understanding healthy development and disease. Therefore, to assess the feasibility of low-field paediatric MR brain imaging, two initial tasks are considered:

– Task 1: Automated quality assurance (QA) to rate the overall quality of low-field MRI images, to ensure that the acquired MR images meet specific standards.
– Task 2: Automatic segmentation of the bilateral hippocampi, due to their importance as a subcortical structure linked to cognitive and memory functions.

Here we explore preliminary results from the above two tasks based on T2 weighted MR images by comparing multiple approaches. Our results show that these images with single modality can provide understanding of overall anatomies and large scale pathologies. However, our results suggest that multiple scan types and quality enhancement approaches might be required for more granular analyses and in-depth exploration of structural development in paediatric populations.

2 Methods

2.1 Classification of Artefacts for Quality Assessment

We aim to perform quality assessment of paediatric brain MR images by assigning scores of 0, 1 and 2 (0 being good quality and 2 meaning there is a very high level of artefacts) across seven artefact domains such as noise, zipper, positioning, banding, motion, contrast and distortion. One of the major challenges is the imbalance in data samples across the different scores: the majority of cases were of good quality and the proportion of data with artefacts was extremely small, especially the ones with a score of 2. Hence, as a first step, we performed appearance-informed transformations to simulate artefacts to augment the data (for classes 1 and 2) for training.

Appearance-Informed Transformations for Simulating Artefacts. We used various appearance-based transformation as specified in Table 1 and a few examples are shown in Fig. 1 for various artefact domains. We used Torch IO library [1] for simulating the artefacts using parameters mentioned in the table. Note that these transformations (specified in Table 1) were used for quality assessment (task 1) alone, since they were specific to the artefacts observed in the data.

Automated Quality Assessment Using Deep Learning. For classification of images based on their quality, we compared the following architectures:

- Multi-headed decoder with single shared encoder [8]
- DenseNet architecture [9]

Of the two architectures, this study is the first ever to experiment with multi-headed decoder framework, to the best of our knowledge, while DenseNet has been already used for similar tasks [2]. The multi-headed decoder was chosen due to its capability to learn common attributes of data due to a shared encoder, which helps in learning perturbations better.

Multi-head Decoder Model for Quality Assessment. Figure 2 shows the network architecture of the multi-head decoder model. The architecture consists of an encoder that is shared across multiple heads of the multiple decoder, that essentially extracts features that are different from high quality images without any artefacts. We used the encoder of a 4-layer deep UNet [16] model, and connected the decoders to the bottleneck of the UNet. Each decoder consisted of 3 fully connected layers (with 4096, 512 and 32 nodes respectively), followed by the output Sigmoid layer with 3 nodes (for classes 0, 1 and 2). The model was trained using an AdamW optimiser with a learning rate of 1×10^{-5} and a patience value of 5. We used a batch size of 8, and train:validation split of 80:20. For standard data augmentations (note that these are separate from the transformations in Sect. 2.1), we used and MONAI transforms for inflating the

Table 1. Appearance-based transformation used for data augmentation (the parameters were selected empirically from held-out data).

artefact domain	Simulation steps	Parameters for steps
Noise (e.g. low-field strength)	Gaussian blur + Blur	**Class 1:** $\sigma_{noise} = [0, 0.1]$, $\sigma_{blur} = [0, 0.6]$ **Class 2:** $\sigma_{noise} = [0, 0.2]$, $\sigma_{blur} = [0, 0.6]$
Zipper (e.g. EM spikes in gradient coil & fluctuating power supply)	Herringbone artefacts + Blur	**Class 1:** No. of spikes = 1, Intensity contrast (between spikes) = [0.1, 0.3], $\sigma_{blur} = [0, 0.6]$ **Class 2:** No. of spikes = 1, Intensity contrast (between spikes) = [0.3, 0.6], $\sigma_{blur} = [0, 0.6]$, $\sigma_{blur} = [0, 0.6]$
Positioning (e.g. discrepancy and phase- and frequency-encoding sampling times)	Translation + Rotation + Blur	**Class 1:** $x_{offset} \in [-10, 10]$ voxels, $y_{offset} \in [-10, 10]$ voxels, $\theta \in [0, 10]$ deg, $\sigma_{blur} = [0, 0.6]$ **Class 2:** $x_{offset} \in [-20, 20]$ voxels, $y_{offset} \in [-20, 20]$ voxels, $\theta \in [10, 30]$ deg, $\sigma_{blur} = [0, 0.6]$
Banding (e.g. due to magnetic field inhomogeneity)	Noise in a random spatial band + Blur	**Class 1:** Same set of parameters as used in noise. **Class 2:** Same set of parameters as used in noise.
Motion (e.g. due to patient movement during scan)	Motion artefacts + Blur	**Class 1:** $\theta \in [-10, 10]$ deg, motion offset $\in [0,3]$ mm, No. of transforms = 2, $\sigma_{blur} = [0, 0.6]$ **Class 2:** $\theta \in [-20, 20]$ deg, motion offset $\in [0.7]$ mm, No. of transforms = 4, $\sigma_{blur} = [0, 0.6]$
Contrast (e.g. due to bias field)	Bias field heterogeneity + Gamma correction + Blur	**Class 1:** Coefficients of polynomial for bias field $\in [0, 0.3]$, order of polynomial = 3, $\gamma \in [0, 0.3]$, $\sigma_{blur} = [0, 0.6]$ **Class 2:** Coefficients of polynomial for bias field $\in [0, 0.6]$, order of polynomial = 5, $\gamma \in [0.3, 0.6]$, $\sigma_{blur} = [0, 0.6]$
Distortion (e.g. from gradient non-linearities and poor shimming)	Elastic deformation + Blur	**Class 1:** Noise parameters are same as those for noise. No. of control points = 7, max. displacement = 9 voxels, Interpolation method = 'bspline', $\sigma_{blur} = [0, 0.6]$ **Class 2:** Noise parameters are same as those for noise. No. of control points = 12, max. displacement = 12 voxels, Interpolation method = 'bspline', $\sigma_{blur} = [0, 0.6]$

training data. We used focal loss (Eq. 1) [14] for training. The focal loss is given by:

$$FL(p_t) = -\alpha_t(1 - p_t)^\gamma log(p_t) \tag{1}$$

where α and γ are weighing and focusing parameters with values of 0.25 and 2 respectively (chosen empirically), and $p_t \in \{0, 1\}$ is the predicted probability. During inference, we applied the trained encoder and individual decoders on each test instances and obtained the predictions for 7 artefact domains and used max-voting for obtaining final artefact prediction.

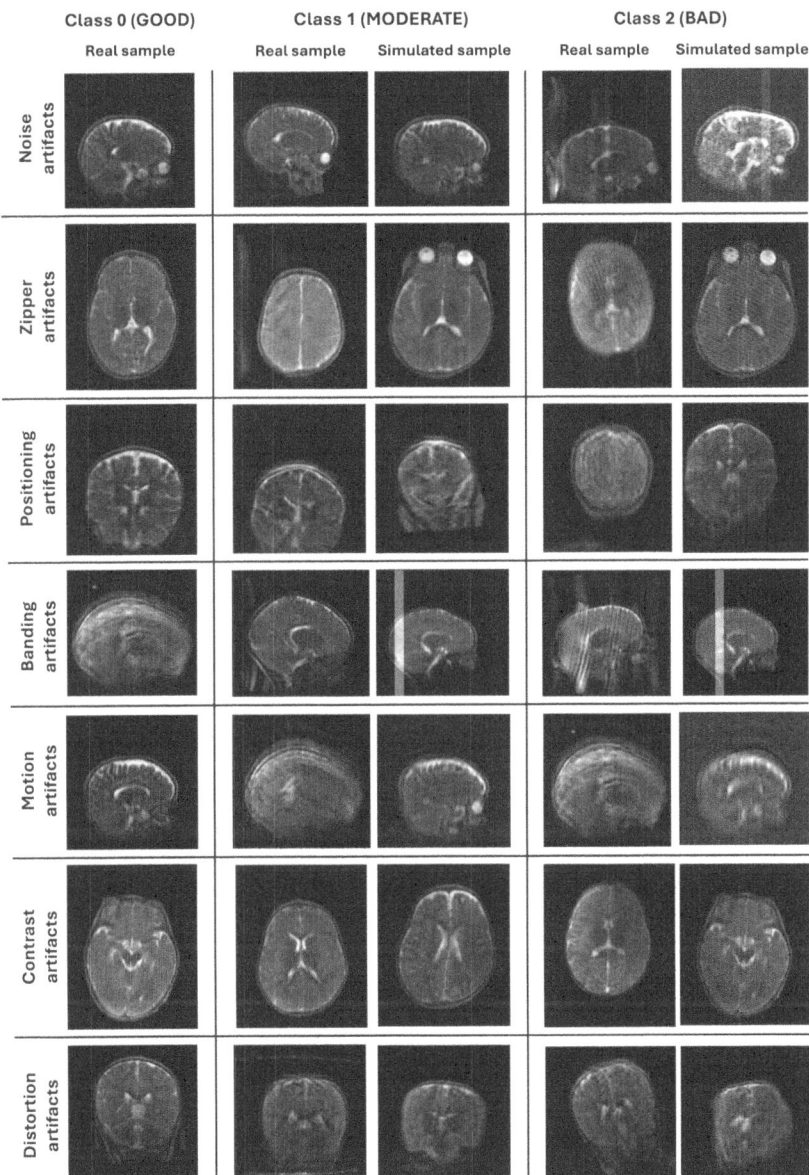

Fig. 1. Simulation of artefacts for various artefact domains for training the quality assessment models.

DenseNet Model Architecture for Quality Assessment. We also trained a DenseNet [9] model using cross-entropy loss for training, individually for each artefact domain. We used the same parameters as used for training the multi-head decoder model.

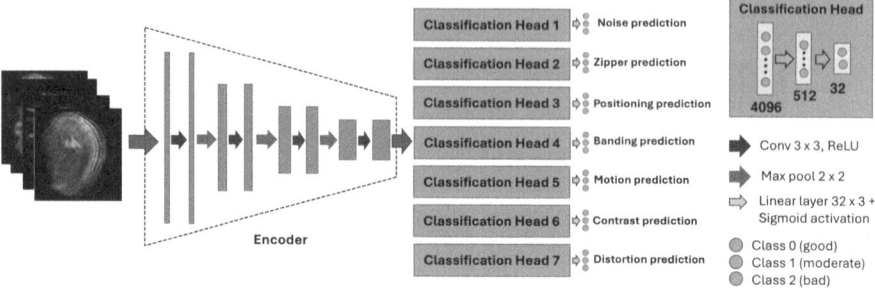

Fig. 2. Architecture of the multi-head decoder model for quality assessment.

2.2 Segmentation of Hippocampus

Preliminary results from a 3D UNet indicated that the low contrast images were causing challenges for the segmentation of the hippocampus, leading to a large degree of under-segmentation. Therefore, we considered a range of different methods based on different paradigms:

- Out-of-the-box Segmentation Approach (FSL FIRST [15])
- Linear Registration of an atlas
- 3d UNet
- 3d UNet + Prior

Out-of-the-Box (OOB) Segmentation Approach (FSL FIRST). The majority of OOB segmentation tools for the hippocampus have been developed for adults and 1.5/3T MR images. Therefore, their expected performance on our images is unknown. We compared results on a subset of the training dataset for the following methods: FSL FIRST [15], Freesurfer [7], SynthSeg [5] and HippoDeep [18]. Only FSL FIRST was able to reliably locate the hippocampus in the images and so is used as the OOB approach.

Linear Registration. For the linear registration approach we aimed to create a study specific hippocampus atlas, that could be then propagated to the individual subjects. We used FSL FLIRT [10] to register all subjects to the first training subject (id: 0001), and then propagated all labels to this subject space and averaged them to produce a study specific hippocampus atlas. Non-linear registration was not used as due to the lack of contrast the algorithm struggled to converge and led to poor registration results. The atlas was then propagated back to the individual subject spaces for the validation samples, and thresholded at 0.1 (chosen empirically from held-out samples) to produce the binary segmentation mask. The average mask can be seen in Fig. 5C.

3D UNet. We considered a Vanilla 3D UNet [16] (64 features at the first level and 4 pooling layers), and trained with a Dice Loss function. Given the images were rigidly aligned, we selected an ROI from the training images centred around

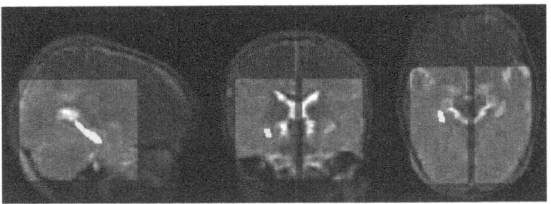

Fig. 3. Selected ROI from T2

each hippocampus separately, as shown in Fig. 3. The images were split at the subject level into training and validation sets, and the model was trained using an AdamW optimiser with learning rate 1×10^{-4} and a patience value of 15. The problem was treated as a binary segmentation task with both hippocampi being represented by the same label value, and standard augmentation was applied throughout training.

3D UNet + Prior. On the held-out validation set, the 3D UNet was seen to have undersegemented the hippocampi, and so a prior was added to encourage the network to create larger segmentations. Specifically, an estimate of the amount of the ROI which should be hippocampus was calculated from the prior produced during the linear registration approach. This was binarised and then the ratio calculated. The model was then trained, inspired by [4], using a loss function that directly aims to match the ratio of background to hippocampus:

$$L_{total} = L_{dice}(\boldsymbol{Y}, \hat{\boldsymbol{Y}}) + \lambda KL(\tau, \hat{\tau}) \qquad (2)$$

where $\boldsymbol{Y}$ is the true segmentation mask, $\hat{\boldsymbol{Y}}$ is the predicted segmentation mask, KL is the KL divergence between the true tissue ratio τ, and the predicted tissue ratio $\hat{\tau}$. λ is the weighting factor between the two loss functions, set to 5 empirically.

Datasets Used. The data used were from the LISA Challenge [13] consisting of low-field MRI dataset of over 300 paediatric T2 scans, acquired using a 0.064T MR Scanner with a sequence type of spin echo, TR 1.5 s, TE 5ms and TI 400ms. The images were split (at the subject level) into training and validation sets with a split of 80:20%. Quality assessment scores (values of 0, 1 and 2) were available across seven artefact domains (noise, zipper, positioning, banding, motion, contrast, distortion) in the CSV file format. Results are reported (Sect. 3.1) on the 14 validation samples available as part of the Challenge. For segmentation task (task 2), the models were trained using the 79 manually labelled T2 MR images with manual labels. No preprocessing was performed other than those detailed in methods. Results are reported on the 12 validation samples, preprocessed identically to the training data where appropriate.

3 Results and Discussion

3.1 Quality Assessment Results

The results of comparison of multi-head decoder with DenseNet model are shown in Table 2. As mentioned earlier, the main challenge in this task is the heavy class imbalance between class 0 and others (especially class 2, which are very low), hence leading the model to be biased towards class 0 and the anatomical structures are not being clearly distinguishable leading to poor predictive quality. However, it can be seen that the performance improved with the addition of simulated data using transformations described Table 1. Between the architectures, DenseNet provided much better performance ($Accuracy_{weighted}$: 0.823), when compared to the multi-head decoder model ($Accuracy_{weighted}$: 0.741), likely due to the ability of the former to extract complex spatial and contextual features that explain the subtle changes in the brain structure.

Table 2. Quality assessment results averaged across the dataset for various evaluation metrics. MHD: Multi-head decoder. WOA: without using augmentations, WA: With augmentations.

Metric	MHD (WOA)	MHD (WA)	DenseNet (WOA)	DenseNet (WA)
$F1_{micro}$	0.357	0.741	0.735	**0.823**
$F1_{macro}$	0.285	0.450	0.444	**0.510**
$F1_{weighted}$	0.450	0.767	0.762	**0.818**
$F2_{micro}$	0.357	0.741	0.735	**0.823**
$F2_{macro}$	0.288	0.472	0.468	**0.514**
$F2_{weighted}$	0.372	0.748	0.742	**0.819**
$Precision_{micro}$	0.357	0.741	0.737	**0.823**
$Precision_{macro}$	0.421	0.510	0.446	**0.554**
$Precision_{weighted}$	0.770	0.819	0.817	**0.834**
$Recall_{micro}$	0.357	0.741	0.735	**0.823**
$Recall_{macro}$	0.443	**0.532**	0.529	0.526
$Recall_{weighted}$	0.357	0.741	0.735	**0.823**
$Accuracy_{micro}$	0.357	0.741	0.735	**0.823**
$Accuracy_{macro}$	0.357	0.741	0.735	**0.823**
$Accuracy_{weighted}$	0.357	0.741	0.735	**0.823**

3.2 Segmentation Results

The results comparing the different segmentation approaches are reported in Table 3. It can be seen that the OOB produced poor quality segmentations for the low-field data, as expected due to the large domain shift between this

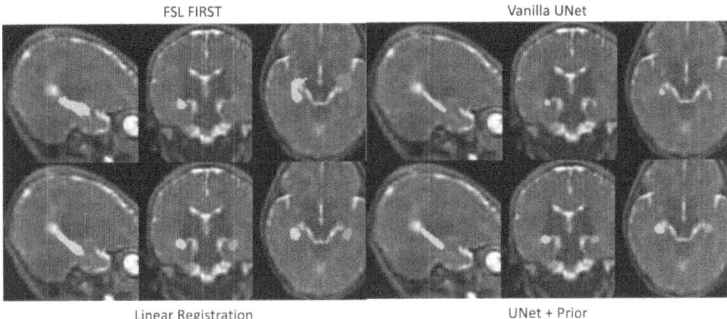

Fig. 4. Validation example segmentation results comparing the result from the four methods.

Table 3. Segmentation results averaged across left and right hippocampi for the five different metrics considered.

Method	Dice	HD	HD95	ASSD	RVE
FSL FIRST	0.20±0.22	53.26±81.28	50.30±82.44	8.34±10.17	0.69±0.47
Linear Registration	**0.61±0.15**	**7.32±8.03**	**3.12±0.89**	**1.03±0.43**	0.37±0.20
3D UNet	0.58±0.18	8.94±8.22	4.44±2.17	1.24±0.94	**0.22±0.14**
3D UNet + Prior	0.56±0.22	11.50±12.51	6.74±8.78	2.46±4.00	11.05±37.56

data and the data FSL FIRST was trained on. The vanilla UNet produced the better results, although far below the performance that would be expected for hippocampal segmentation on adult high field data. The addition of the prior reduced the segmentation quality slightly but the performance was very asymmetric (left 0.52±0.28, right 0.59±0.19), likely due the significantly different sizes of the manual masks for each hemisphere (left volume: 1160 ± 308, right volume: 1225 ± 338, paired ttest p=0.0008)). Registration of the average hippocampal volume provided the best results across the metrics apart from relative volume error, probably indicating that the atlas over estimates the size of the hippocampus. This demonstrates the lack of signal available in the images, as even very simple DL-based approaches would be expected to out-perform registration based approaches.

Figure 4 shows an example segmentation from the validation data (no manual label available), comparing the four approaches. It can clearly be seen that FIRST over segments the hippocampus whereas the 3D UNet produces the lowest volume segmentations. The addition of the prior clearly increases the size of the segmented region. The localisation of the hippocampus is consistent across the linear registration, Vanilla 3D UNet and 3D UNet + prior approaches.

Training and evaluation of the DL-based models were affected by the quality of the manual masks: unsurprisingly given the low contrast and relatively poor image quality, the quality of the training masks varied, biasing results. For

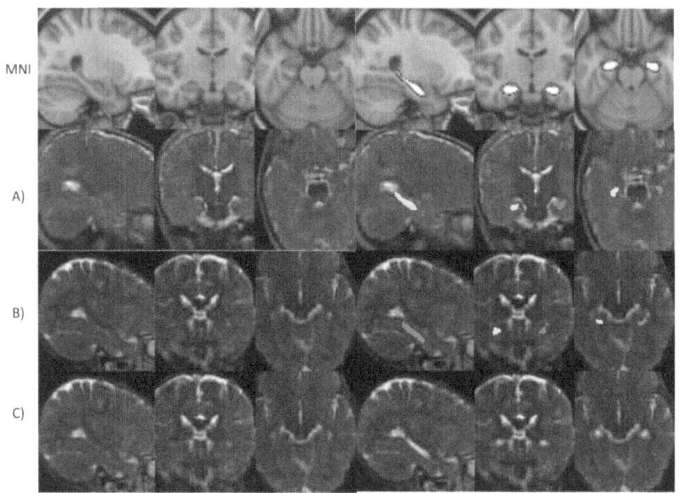

Fig. 5. Example hippocampal segmentations. Row 1: MNI T1 2mm template with the Harvard-Oxford probabilistic atlas, thresholded at 0.5. A): Example A showing a sample the 3D UNet repeatedly performed well (DSC > 0.6) on with the tail of the atlas touching the ventricle. B) Example B showing a sample where the 3D UNet performed poorly (DSC > 0.1), where the localisation of the hippocampus in the target appears different to the MNI template and sample A, with the label not reaching the ventricles. C) Example B with the average hippocampus mask registered to it, showing that the average hippocampus mask is located disjoint to the manual segmentation mask.

instance, some samples appear to have labels which mislocalise the hippocampus (Fig. 5) where it can be seen that the example B's mask appears to be shifted lower than the expected location when considering the other examples and the average hippocampus mask registered to that subject.

4 Conclusions

In this work, we provided preliminary analysis on low-field paediatric brain for two tasks: quality assessment and hippocampal segmentation, by comparing multiple approaches. Our results for quality assessment show that simulated artefacts helped to counteract the class imbalance while a densely connected architecture provided 10% increase in accuracy. For hippocampal segmentation, we obtained the best results with registration of the study specific average, surpassing out-of-box methods which originally developed mainly for adult brains and deep learning approaches. As future work, more complicated DL models (e.g. transformer-based networks) would be implemented using self-supervised learning approach for increasing the performance. However, a vast amount of improvement would be required before the segmentation outputs can be used for automated granular analyses of paediatric brains. The Python implementation of our code is publicly available at https://github.com/v-sundaresan/LISA2024_QA.

Acknowledgments. This work was supported by DBT/Wellcome Trust India Alliance Fellowship [grant number IA/E/22/1/506763]. This work was also supported in part by Start-up Research Grant [grant number SRG/2023/001406] from the Science and Engineering Research Board, India and in part by Siemens Healthineers-CDS Collaborative Laboratory of Artificial Intelligence in Precision Medicine. VS is also supported by Pratiksha Trust, Bangalore, India [grant number FG/PTCH-23-1004] and the Seed Research Grant [grant number IE/RERE-22-0583] from the Indian Institute of Science, Bangalore, India.

Disclosure of Interests. The authors have no competing interests to declare that are relevant to the content of this article.

References

1. Torchio: a python library for efficient loading, preprocessing, augmentation and patch-based sampling of medical images in deep learning. Comput. Methods Programs Biomed. (2021)
2. Ahmad, A., Parker, D., Dheer, S., Samani, Z.R., Verma, R.: 3D-QCNet - a pipeline for automated artifact detection in diffusion MRI images. Comput. Med. Imaging Graph. **103**, 102151 (2023). https://doi.org/10.1016/j.compmedimag.2022.102151
3. Balboni, E., et al.: The impact of transfer learning on 3D deep learning convolutional neural network segmentation of the hippocampus in mild cognitive impairment and alzheimer disease subjects. Hum. Brain Mapp. **43**(11), 3427–3438 (2022). https://doi.org/10.1002/hbm.25858, https://onlinelibrary.wiley.com/doi/abs/10.1002/hbm.25858
4. Bateson, M., Kervadec, H., Dolz, J., Lombaert, H., Ben Ayed, I.: Source-free domain adaptation for image segmentation. Med. Image Anal. **82**, 102617 (2022). https://doi.org/10.1016/j.media.2022.102617
5. Billot, B., et al.: SynthSeg: segmentation of brain MRI scans of any contrast and resolution without retraining **86**, 102789 (2023). https://doi.org/10.1016/j.media.2023.102789, https://www.sciencedirect.com/science/article/pii/S1361841523000506
6. Dinsdale, N.K., Bluemke, E., Sundaresan, V., Jenkinson, M., Smith, S., Namburete, A.I.L.: Challenges for machine learning in clinical translation of big data imaging studies. Neuron **110**, 3866–3881 (2021). https://api.semanticscholar.org/CorpusID:235829550
7. Fischl, B., et al.: Whole brain segmentation: automated labeling of neuroanatomical structures in the human brain. Neuron **33**(3), 341–355 (2002). https://doi.org/10.1016/s0896-6273(02)00569-x
8. Hafiz, A.M., Bhat, R.U.A., Parah, S.A., Hassaballah, M.: Se-md: a single-encoder multiple-decoder deep network for point cloud reconstruction from 2D images. Pattern Anal. Appl. **26**(3), 1291–1302 (2023)
9. Huang, G., Liu, Z., Van Der Maaten, L., Weinberger, K.Q.: Densely connected convolutional networks. In: Proceedings of the IEEE Conference on Computer Vision and Pattern Recognition, pp. 4700–4708 (2017)
10. Jenkinson, M., Bannister, P., Brady, M., Smith, S.: Improved optimization for the robust and accurate linear registration and motion correction of brain images. Neuroimage **17**(2), 825–841 (2002). https://doi.org/10.1006/nimg.2002.1132, https://www.sciencedirect.com/science/article/pii/S1053811902911328

11. KS, A., V, D.: Hippocampus in health and disease: an overview. Indian Acad Neurol. (2012)
12. LeCun, Y., Bengio, Y., Hinton, G.: Deep learning. Nature **521**(7553), 436–444 (2015)
13. Lepore, N., et al.: Low field pediatric brain magnetic resonance image segmentation and quality assurance. In: 27th International Conference on Medical Image Computing and Computer Assisted Intervention (MICCAI 2024). Zenodo (2024). https://doi.org/10.5281/zenodo.10992222
14. Lin, T.Y., Goyal, P., Girshick, R., He, K., Dollár, P.: Focal loss for dense object detection. In: Proceedings of the IEEE International Conference on Computer Vision, pp. 2980–2988 (2017)
15. Patenaude, B., Smith, S.M., Kennedy, D.N., Jenkinson, M.: A bayesian model of shape and appearance for subcortical brain segmentation. Neuroimage **56**(3), 907–922 (2011). https://doi.org/10.1016/j.neuroimage.2011.02.046, https://www.sciencedirect.com/science/article/pii/S1053811911002023
16. Ronneberger, O., Fischer, P., Brox, T.: U-net: convolutional networks for biomedical image segmentation. In: Medical Image Computing and Computer-Assisted Intervention – MICCAI 2015 (2015)
17. Sudlow, C., et al.: UK biobank: an open access resource for identifying the causes of a wide range of complex diseases of middle and old age. PLOS Med. **12**(3) (2015). https://doi.org/10.1371/journal.pmed.1001779
18. Thyreau, B., Sato, K., Fukuda, H., Taki, Y.: Segmentation of the hippocampus by transferring algorithmic knowledge for large cohort processing. Med. Image Anal. **43**, 214–228 (2018). https://doi.org/10.1016/j.media.2017.11.004, https://www.sciencedirect.com/science/article/pii/S1361841517301597

Open Access This chapter is licensed under the terms of the Creative Commons Attribution 4.0 International License (http://creativecommons.org/licenses/by/4.0/), which permits use, sharing, adaptation, distribution and reproduction in any medium or format, as long as you give appropriate credit to the original author(s) and the source, provide a link to the Creative Commons license and indicate if changes were made.

The images or other third party material in this chapter are included in the chapter's Creative Commons license, unless indicated otherwise in a credit line to the material. If material is not included in the chapter's Creative Commons license and your intended use is not permitted by statutory regulation or exceeds the permitted use, you will need to obtain permission directly from the copyright holder.

Axis-Guided Quality Assessment and Multi-label Hippocampal and Ventricular Segmentation in Low-Resolution Pediatric Brain MRI

Hyunwook Kim[1], Jinew Seo[1], Seiyoung Ryu[1], Joon hyung Park[1], Sungchul On[1,2], and Jinwha Choi[1](✉)

[1] Department of Convergence Medicine, Asan Medical Institute of Convergence Science and Technology, University of Ulsan College of Medicine, Asan Medical Center, 388-1 Pungnap2-Dong, Songpa-Gu, Seoul, Republic of Korea
jazzina7@gmail.com

[2] Department of Biomedical Engineering, BK21 Project, Asan Medical Center, University of Ulsan College of Medicine, Seoul, Republic of Korea

Abstract. The Swoop system of Hyperfine Inc. is an affordable, ultra-low-field MRI developed for use in a clinical setting. However, despite its advantages, the relatively low resolution of 64mT MRI data poses additional challenges, especially in examining small structures such as the hippocampus or vessels. As a part of our attempt at the Low field pediatric brain magnetic resonance Image Segmentation and Quality Assurance (LISA) Challenge 2024, we developed two deep learning-based models. First, to evaluate the image quality of 64mT T2 brain MRI data, we implemented an axis classifier module to improve the model performance. Second, for segmentation of the hippocampus in the MRI, a multi-label learning method was used for more accurate segmentation. With these models, we expect to alleviate the accessibility barrier to brain MRI.

Keywords: Swoop system · MRI · Deep Learning · Hippocampus

1 Introduction

The importance of Magnetic Resonance Imaging (MRI) as a medical radiology modality has constantly increased, as it plays an integral part in diagnosis and management of various disease entities including stroke [1]. However, conventional MRIs commonly used in clinical settings have significant accessibility issues due to their high costs. Due to their high magnetic strengths (1.5-3T), they also require meticulously tuned operational settings, such as a shielded room for safe and effective use. As a result, there is a severe discrepancy in MRI accessibility in communities, and the World Health Organization report from 2008 showed that about 90% of the world population did not have access

---

Supplementary Information The online version contains supplementary material available at https://doi.org/10.1007/978-3-031-83008-2_5.

© The Author(s) 2025
N. Lepore and M. G. Linguraru (Eds.): LISA 2024, LNCS 15515, pp. 53–62, 2025.
https://doi.org/10.1007/978-3-031-83008-2_5

to MRI services. Additionally, conventional high-field MRI, while providing detailed images, is often not feasible for young children due to the need for sedation and associated risks.

To address these limitations, Hyperfine developed the Swoop system, an ultra-low-field MRI, as an alternative option. Because the swoop system operates in substantially low magnetic field strength (64mT), it does not require a shielded room. It is also built into a portable platform, enabling the technicians to acquire the data directly from the bedside. Additionally, several studies have demonstrated that 64mT MRI, when enhanced with deep learning, is clinically as effective in diagnostics as conventional MRI [2]. Arnold et al. [3] compared the diagnostic performance of 3T and 64mT images by transforming 3T images into matched 64mT images. They used a Convolutional Neural Network to detect pathology patterns on both the 3T images and transformed 64mT images, and they showed comparable results for macro-scale pathology (i.e., gliomas and medium-large vessel strokes). Although its diagnostic performance is not the same as that of conventional MRI, its portability and cost-effectiveness make the Swoop system a potent alternative to traditional MRIs, especially in clinical settings that cannot meet the complex operational requirements of conventional MRIs.

The diagnostic value of brain MRI depends not only on the MRI machine itself but also on the quality of the acquired images. Skilled technicians or radiologists often ensure that image quality meets diagnostic standards. A part of MRI quality assessment is minimizing the distortion caused by various artifacts. Some common artifacts in MRI images include blurring, inappropriate contrast, noises, and ghost images [4]. Recently, attempts have been made to build a deep learning-based automated model to assess such artifacts. Risager et al. developed a 3D ResNet-50 model that detects the presence of 6 artifacts: contrast changes, bias field, Gibbs ringing, motion ghosting, Rician noise, and blur effect [5]. Although their model reliably detected the presence of artifacts, models were trained and explicitly tested with 3T brain MRI data of adults. A pediatric brain undergoes constant structural changes until age 6, when the brain reaches approximately 95% of its total size. It also has structural characteristics such as a less pronounced distinction between gray matter and white matter, making its MRI presentation very different from that of an adult brain. Hence, a deep learning-based model may show different performance on pediatric brain MRIs, and lower resolution of MRI images may also affect the model's performance [6]. For the LISA challenge 2024, the first part of our goal was to build a robust model that could detect distortions in 64mT pediatric brain MRI.

The hippocampus plays a crucial role in cognitive development and learning during early childhood. It is essential for memory formation, spatial navigation, and emotional regulation, all of which are foundational for the overall cognitive abilities of a developing child. When there are issues in the development or function of the hippocampus, it can lead to developmental delays and cognitive impairments. These impairments are particularly concerning in the context of increasing prevalence rates of neurodevelopmental disorders such as autism spectrum disorder (ASD) and attention deficit hyperactivity disorder (ADHD), which have been linked to hippocampal dysfunction [6]. The rising incidence of these disorders has not only increased the burden on healthcare systems due to higher medical costs. Still, it has also underscored the need for early diagnosis and intervention.

Given the importance of the hippocampus in these neurodevelopmental disorders, numerous studies have focused on examining its structure and function using brain MRI. These studies aim to better understand the pathophysiology of ASD, ADHD, and other pediatric conditions. However, as previously mentioned, technical and accessibility issues of conventional MRI limit our ability to accurately study hippocampal abnormalities in these critical years of development. Although ultra-low field MRIs such as Hyperfine Swoop successfully address these issues, their low resolution presents a significant challenge in studying small brain structures like the hippocampus [3]. Accurate segmentation is crucial for assessing hippocampal volume and morphology, which are critical indicators of neurodevelopmental health.

This study seeks to bridge this gap by developing an automatic segmentation method for the hippocampus in early childhood brain MRI scans obtained with a low-field 64mT system. This research aims to advance the field by providing a robust deep learning-based solution that can accurately segment the hippocampus despite the inherent limitations of low-field MRI. This approach not only holds the potential for improving diagnostic accuracy in resource-limited settings but also paves the way for broader applications of low-field MRI in pediatric neuroimaging.

2 Methods

2.1 Data Collection

Quality Assessment
The organizers provided 474 pediatric brain MRI data of 158 individuals. For each individual, there were three brain MRI data for each of three anatomical planes: axial, coronal, and sagittal. Along with MRI data, CSV files containing quality assessment scores for seven artifact domains were provided (Fig. 1).

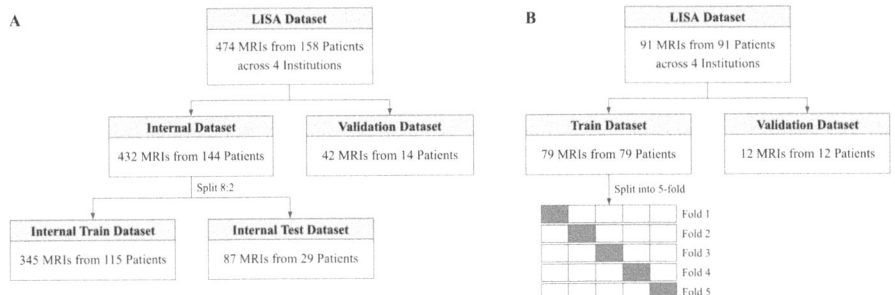

Fig. 1. Data collection and partitioning. A: Datasets for 64mT MRI quality assessment model. B: Datasets for hippocampus segmentation model.

Hippocampal Segmentation

Seventy-nine pairs of high-field and low-field MRI data were provided for training purposes. Bilateral hippocampal segmentation data were given in NIFTI format and aligned with the high-field scan space. Segmentation was manually performed using high-field T2 MRI images (3T or 1.5T). The segmentation protocol, including the corrections, was reviewed and approved by a board-certified pediatric neuroradiologist with several decades of experience. Skilled technicians acquired all the data for both tasks in the following institutions: (1) Kawempe National Referral Hospital, Makerere University, Kampala, Uganda; (2) CUBIC, University of Cape Town, South Africa; and (3) Warren Alpert Medical School at Brown University, Providence, RI, USA, and the Advanced Baby Imaging Lab, Rhode Island Hospital, Providence, RI, USA. According to the organizers, the 64mT MRI data were acquired using Hyperfine Swoop with a sequence type of spin echo, repetition time (TR) 1.5s, echo time (TE) 5ms, and inversion time (TI) 400ms. All MRI data were evaluated by skilled image evaluators in corresponding institutions.

2.2 Target Artifacts for Quality Assessment

For image quality assessment, the six common MRI artifacts were selected by the organizers: motion, contrast, noise, zipper, positioning, and banding. Bottani et al. [7] suggested that detecting motion, contrast, and noise artifacts is adequate for automated quality control of brain MRI. In addition to these artifacts, zipper, positioning, banding, and distortion were added by the organizers to assess ultra-low field Hyperfine Swoop MRI. The extent to which each artifact affects the image quality is graded from 0 to 2. The instruction paper did not specify the definition and scoring for distortion [7].

2.3 Additional Labeling for Segmentation

Since the MR images were acquired using a low-field 64mT system, segmentation of the hippocampus remains challenging. We trained the model using the anatomical structure to improve segmentation performance. Lateral ventricles are bilateral cavities filled with cerebrospinal fluid and thus show hyperintensity in T2 MRI, which allows clear distinction among the ventricles and adjacent regions, including the hippocampus. Multi-label learning in our model combines the ventricle and hippocampus labels using the anatomical structure. Two experienced labelers manually segmented the lateral ventricles on the image and added each label for the left and right lateral ventricles.

2.4 Model Architecture

Quality Assessment

We developed a model based on the DenseNet [8] architecture for the automated quality assessment task. It is particularly effective in medical image analysis, especially for MRI data, due to its densely connected layers and skip connections that enhance feature propagation and reuse [8]. Since each patient in our dataset underwent three brain MRI scans with different anatomical orientation focuses, we incorporated an axis classifier module into the model. This module accurately predicts the focused orientation of each scan (coronal, axial, sagittal), allowing the model to account for variations in image shape and orientation (Fig. 2).

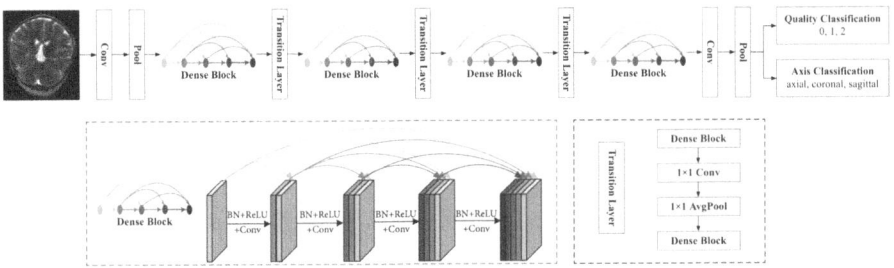

Fig. 2. Axis-guided classification model based on DenseNet [8]

Hippocampal Segmentation
We built a model based on nnU-Net [9] for the bilateral hippocampal segmentation task. nnU-Net is a framework built on top of the U-Net architecture designed for medical image segmentation. The model comprises an encoder-decoder network with skip connections that capture multi-scale features. nnU-Net learns by processing patches of the input images rather than the entire image at once. During training, the input images are divided into smaller patches, which allows nnU-Net to efficiently handle large medical images, manage memory usage, and focus on relevant details within each patch (Fig. 3).

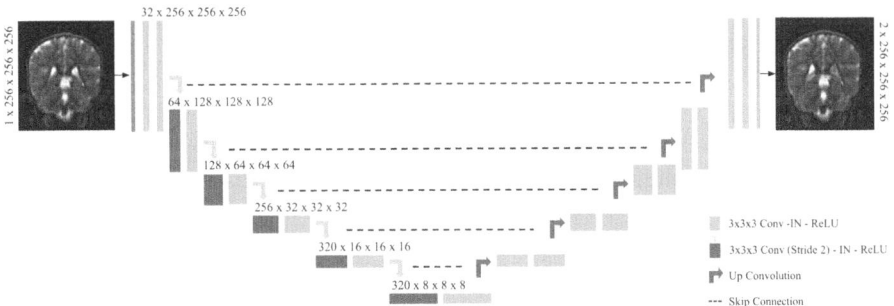

Fig. 3. Hippocampal and ventricular segmentation model based on 3D nnU-Net

2.5 Training Configurations

Quality Assessment
The training was performed on an NVIDIA RTX A6000 48GB, using 100 epochs with a batch size of 4. The Adam optimizer was employed with a learning rate of 1e-5 alongside a Cosine Annealing Learning Rate Scheduler. Cross-entropy loss was used for both

quality assessment and axis classification, with equal weighting on the combination. Preprocessing involved Z-score normalization, and data augmentation was limited to rotation along the oriented axis to prevent any impact on artifact grading. The total training time averaged 4 h 30 min, and the model's performance was evaluated using F1-score, F2-score, precision, recall, and accuracy.

Hippocampal Segmentation
The training was conducted on an NVIDIA Quadro RTX 8000 D6 48GB GPU. The training process included 500 epochs per phase, utilizing a batch size of 2. The optimizer was Stochastic Gradient Descent, configured with a learning rate of 0.01, a momentum of 0.99, and a weight decay of 0.0005. Preprocessing of the images involved Z-score normalization. Data augmentation was applied through scaling and rotation along each axis. Horizontal flip was not used because the left and right hippocampus labels are divided. The model utilized Dice focal loss for the loss function. The weights for each loss were combined in a ratio of 1:1. The total training time was 11 h and 30 min on average per fold, and training was performed in parallel for five folds. After training on all five folds, one output label was derived for each model through the ensemble. Binary closing was performed twice after the prediction for post-processing. For the evaluation, Dice similarity coefficients (DSC), Hausdorff distance (HD), Hausdorff distance 95% (HD95), average symmetric surface distance (ASSD), and relative volume error (RVE).

3 Results

3.1 Quality Assessment

Table 1 shows the performance of three DenseNet configurations: baseline, with axis rotation, and with both axis rotation and axis classification. Across all metrics, the model with both axis rotation and classification performed the best.

For micro metrics, all values improved uniformly from 0.823 in the baseline model to 0.847 with axis rotation and 0.854 with axis classification. In macro metrics, F1-score improved from 0.461 to 0.524 with axis rotation, and 0.531 with the addition of axis classification, while F2-score increased from 0.468 to 0.535 and 0.536, respectively. Precision and Recall similarly improved, with the final model achieving 0.588 in Precision and 0.556 in Recall. In weighted metrics, F1-score, F2-score, Precision, and Recall increased from 0.824, 0.823, 0.827, and 0.823 to 0.846, 0.849, 0.858, and 0.854, respectively, showing consistent improvement across all metrics.

3.2 Hippocampal Segmentation

Table 2 shows the evaluation results of the predicted segmentations using the model with only the hippocampus label. The single-label model performed best at fold 2 with a Dice score of 0.70 and a Hausdorff distance of 5.89 mm. Table 3 shows the evaluation results of the predicted segmentations using the model with multi-label using the ventricle. The multi-label model showed the best performance overall with a Dice score of 0.70 and a Hausdorff distance of 5.89 mm predicted from 5 folds.

Table 1. Comparison of quality assessment for internal validation sets among various models

Micro						
Axis Rotation	Axis Classification	F1-score	F2-score	Precision	Recall	Accuracy
X	X	0.823	0.823	0.823	0.823	0.823
O	X	0.847	0.847	0.847	0.847	0.847
O	O	0.854	0.854	0.854	0.854	0.854
Macro						
Axis Rotation	Axis Classification	F1-score	F2-score	Precision	Recall	Accuracy
X	X	0.461	0.468	0.462	0.487	0.823
O	X	0.524	0.535	0.533	0.551	0.847
O	O	0.531	0.536	0.588	0.556	0.854
Weighted						
Axis Rotation	Axis Classification	F1-score	F2-score	Precision	Recall	Accuracy
X	X	0.824	0.823	0.827	0.823	0.823
O	X	0.839	0.843	0.841	0.847	0.847
O	O	0.846	0.849	0.858	0.854	0.854

Table 2. Evaluation results of hippocampal segmentation for the single-label model.

	DSC	HD (mm)	HD95 (mm)	ASSD (mm)	RVE
Fold 0	0.64 ± 0.27	9.32 ± 13.92	5.64 ± 12.34	3.92 ± 10.93	0.15 ± 0.09
Fold 1	0.70 ± 0.19	5.89 ± 8.05	2.20 ± 1.79	0.80 ± 0.99	0.15 ± 0.09
Fold 2	0.69 ± 0.21	8.41 ± 11.43	3.47 ± 4.76	0.87 ± 0.95	0.16 ± 0.11
Fold 3	0.65 ± 0.26	8.79 ± 12.36	5.07 ± 10.34	3.32 ± 8.95	0.16 ± 0.10
Fold 4	0.68 ± 0.20	26.05 ± 62.97	22.60 ± 63.62	0.84 ± 0.77	0.21 ± 0.19

* DSC (Dice Score Coefficient), HD (Hausdorff Distance), HD95 (Hausdorff Distance 95%), ASSD (Average Symmetric Surface Distance), RVE (Relative Volume Error)

Table 3. Evaluation results of hippocampal segmentation for the multi-label model.

	DSC	HD (mm)	HD95 (mm)	ASSD (mm)	RVE
Fold 0	0.70 ± 0.19	5.64 ± 7.91	1.97 ± 1.45	0.73 ± 0.85	0.16 ± 0.07
Fold 1	0.70 ± 0.17	6.04 ± 8.15	2.21 ± 1.49	0.80 ± 0.89	0.16 ± 0.09
Fold 2	0.70 ± 0.17	5.62 ± 7.96	1.96 ± 1.38	0.74 ± 0.84	0.17 ± 0.08
Fold 3	0.71 ± 0.19	5.39 ± 7.89	1.90 ± 1.32	0.69 ± 0.71	0.12 ± 0.06
Fold 4	0.70 ± 0.18	5.69 ± 7.77	2.02 ± 1.48	0.76 ± 0.89	0.14 ± 0.10
Ensemble	0.72 ± 0.19	5.66 ± 7.78	1.92 ± 1.51	0.71 ± 0.83	0.13 ± 0.10

* DSC (Dice Score Coefficient), HD (Hausdorff Distance), HD95 (Hausdorff Distance 95%), ASSD (Average Symmetric Surface Distance), RVE (Relative Volume Error)

4 Discussion

The ablation study of automated quality assessment models shows that incorporating axis rotation and axis classification into the DenseNet model significantly improves its performance. Comparison of our two enhanced models, one with only axis rotation and the other with both axis rotation and axis classification, to the baseline DenseNet model show consistent improvements across all evaluation metrics. Axis rotation improves the model's ability to generalize across different image orientations, which is crucial for detecting artifacts in pediatric brain MRIs because orientation differences can affect detection accuracy. Adding axis classification further refines the model's predictions by enabling it to recognize specific anatomical planes. This combination enhances artifact detection and increases robustness, making the model more reliable for quality assessment, particularly valuable in low-field clinical settings for effective diagnosis.

In our hippocampal segmentation study, a multi-label model using the adjacent anatomical structure, the lateral ventricle, showed superior results to a model that only used hippocampal labels. The total dice results improved from 67% to 72%, 20 mm on some folds to 5.66 mm for HD, 7.8 mm to 1.92 mm for HD95, from 1.95 mm to 0.71 mm for ASSD, and from 17% to 13% for RVE. Segmentation performance was generally better for the right hippocampus than the left in all evaluation results. This asymmetry in performance might be attributed to the inherent anatomical differences between the two hemispheres [10, 11], challenges posed by low-field MRI's reduced resolution and contrast, or some errors that occurred during label registration from the high-field MR image.

In adult brain MRI studies, hippocampal segmentation using deep learning methods has typically achieved higher accuracy than in pediatric studies [6]. A notable study by Hazarika et al. [12] utilizing the modified U-Net architecture reported a Dice coefficient of approximately 0.97 for adult hippocampal segmentation on a 3T MRI dataset. This performance is significantly higher than the results observed in our study. The discrepancy can be attributed to the differences in MRI resolution and the challenges posed by the smaller size and less distinct boundaries of the early childhood hippocampus.

One of the main strengths of this study is the focus on low-field MRI, an accessible and cost-effective 64 mT Hyperfine scanner. As the first study of automated hippocampus segmentation in pediatric low-resolution brain MRI images, this approach addresses a significant gap in the current literature, where most hippocampal segmentation studies are based on high-field MRI.

There are still some limitations. For the quality assessment, there was a variability in the magnitude of artifacts within the same score category, which made it challenging for the model to accurately classify them. For instance, images with a noise grade of 2 displayed a range of features, complicating the classification process. Additionally, the dataset had a significant class imbalance, especially with the banding artifact, where only one image was assigned a grade of 2 across the entire training set, hindering effective model learning. The overall quantity of data was limited, and attempts to address this through upsampling were largely ineffective, with augmentation being the only strategy that provided some benefit. Furthermore, the subjective nature of artifact grading could lead to inconsistencies in labels, as different graders may interpret artifact severity differently, thereby introducing noise into the training data.

For the hippocampal segmentation, the hippocampus label was segmented in a high-resolution image, then aligned to a low-resolution image, and compared with the results, so there may be differences from the actual location of the hippocampus in the low-resolution image. Second, the amount of data is minimal, so more advanced performance can be expected if additional data is secured and training is performed. Third, the accuracy of left hippocampal segmentation was lower, which suggests potential areas for improvement in the algorithm. Expanding the dataset to include more validation and test images from diverse populations could also improve the model's robustness and generalizability.

For a low-field MRI system to have a clinical value, it is critical that the acquired MRI data are of adequate quality. Our first model allows the clinicians to have a better control over the quality of images. Additionally, our segmentation model, with its improved performance, could significantly enhance diagnostic capabilities of the low field MRI. Conventional high-field MRI systems are expensive, require substantial maintenance, and are often inaccessible in low-income regions. Accurate quality assessment and hippocampal segmentation for the low field MRI would greatly improve its functionality as a possible alternative to conventional MRIs in resource limited settings, aiding early diagnosis and intervention for neurodevelopmental disorders such as ASD and ADHD.

5 Conclusion

This study successfully developed an automatic segmentation method for classifying MR image quality assessment and the bilateral hippocampi in pediatric brain MRI scans acquired with a 64mT MRI system. This low-resolution Brain MRI hippocampus segmentation study is an important starting point for future pediatric Brain MRI image analysis studies. It will provide an essential framework for research on pediatric developmental disorders.

Data and Code Availability. The study's dataset including the manually segmented ventricle data, could be available from the first author and the host of the LISA 2024 challenge at a reasonable request. The codes used in the studies are available at the following GitHub repositories: https://github.com/wooks527/lisa2024-task1-qa, https://github.com/wooks527/lisa2024-task2-seg.

References

1. van Beek, E.J.R., et al.: Value of MRI in medicine: more than just another test?" J. Magn. Reson. Imaging **49**(7), e14–25 (2019). Wiley Online Library https://doi.org/10.1002/jmri.26211
2. Mazurek, M.H., et al.: Portable, bedside, low-field magnetic resonance imaging for evaluation of intracerebral hemorrhage. Nat. Commun. **12**(1), 5119 (2021). www.nature.com, https://doi.org/10.1038/s41467-021-25441-6
3. Arnold, T.C., et al.: Simulated diagnostic performance of low-field MRI: harnessing open-access datasets to evaluate novel devices. Magn. Reson. Imaging **87**, 67–76 (2022). PubMed Central, https://doi.org/10.1016/j.mri.2021.12.007
4. Price, R., et al.: MR Quality Control Manual. American College of Radiology (2015). www.acr.org/-/media/ACR/Files/Clinical-Resources/QC-Manuals/MR_QCManual.pdf. Accessed 16 Aug 2024

5. Risager, K.V.E., et al.: Non-reference quality assessment for medical imaging: application to synthetic brain MRIs. arXiv arXiv:2407.14994 (2024). http://arxiv.org/abs/2407.14994
6. Herten, A., et al.: Accuracy and bias of automatic hippocampal segmentation in children and adolescents. Brain Struct. Funct. **224**(2), 795–810 (2019). https://doi.org/10.1007/s00429-018-1802-2
7. Bottani, S., et al.: Automatic quality control of brain T1-weighted magnetic resonance images for a clinical data warehouse. Med. Image Anal. **75**, 102219 (2022). https://doi.org/10.1016/j.media.2021.102219
8. Huang, G., et al.: Densely connected convolutional networks. In: IEEE Conference on Computer Vision and Pattern Recognition (CVPR), vol. 2017, pp. 2261–2269. IEEE Xplore (2017). https://doi.org/10.1109/CVPR.2017.243
9. Isensee, F., et al.: nnU-Net: a self-configuring method for deep learning-based biomedical image segmentation. Nat. Methods **18**(2), 203–11 (2021). www.nature.com, https://doi.org/10.1038/s41592-020-01008-z
10. Rogers, B.P., et al.: Systematic error in hippocampal volume asymmetry measurement is minimal with a manual segmentation protocol. Fron. Neuroscience **6**, 179 (2012). https://doi.org/10.3389/fnins.2012.00179
11. Thompson, D.K., et al.: MR determined hippocampal asymmetry in full term and preterm neonates. Hippocampus **19**(2), 118–23 (2009). https://doi.org/10.1002/hipo.20492
12. Hazarika, R.A., et al.: Hippocampus segmentation using U-Net convolutional network from brain magnetic resonance imaging (MRI). J. Digit. Imaging **35**(4), 893–909 (2022). https://doi.org/10.1007/s10278-022-00613-y

Open Access This chapter is licensed under the terms of the Creative Commons Attribution 4.0 International License (http://creativecommons.org/licenses/by/4.0/), which permits use, sharing, adaptation, distribution and reproduction in any medium or format, as long as you give appropriate credit to the original author(s) and the source, provide a link to the Creative Commons license and indicate if changes were made.

The images or other third party material in this chapter are included in the chapter's Creative Commons license, unless indicated otherwise in a credit line to the material. If material is not included in the chapter's Creative Commons license and your intended use is not permitted by statutory regulation or exceeds the permitted use, you will need to obtain permission directly from the copyright holder.

Quality Assurance and Hippocampal Segmentation on Low-Field Pediatric Magnetic Resonance Images

Austin Tapp[1], Rahimeh Rouhi[2,3], Jeffrey Tanedo[2,3], Shreyash Zanjal[2,3], Sean Deoni[4], Marius George Linguraru[1,5], and Natasha Lepore[2,3(✉)]

[1] Sheikh Zayed Institute for Pediatric Surgical Innovation, Children's National Hospital, Washington, DC 20010, USA
{atapp,MLingura}@childrensnational.org
[2] CIBORG Lab, Department of Radiology, Children's Hospital Los Angeles, Los Angeles, CA 90027, USA
[3] Departments of Pediatrics and Biomedical Engineering, University of Southern California, Los Angeles, CA 90089, USA
{rrouhi,nlepore}@chla.usc.edu, {jtanedo,zanjal}@usc.edu
[4] Bill and Melinda Gates Foundation, PO Box 23350, Seattle, WA 98102, USA
Sean.Deoni@gatesfoundation.org
[5] School of Medicine and Health Sciences, George Washington University, Washington, DC 20052, USA

Abstract. Portable ultra-low-field (uLF, i.e., 0.064T) magnetic resonance imaging (MRI) offers a solution to scarce radiological alternatives of resource-limited regions; however, in such regions, MRI system operators and radiologists are novices to the underrepresented modality. Therefore, automatic methods that confirm image acquisition of appropriate quality for diagnosis and segment and measure critical anatomical structures are required to support rural sites. This paper describes our approach to two tasks presented in the LISA 2024 Challenge: (1) quality assurance of pediatric low-field MRI data using DenseNet, and (2) segmentation of the bilateral hippocampi using nnUNet. As uLF MRI natively introduces unique image quality challenges, we trained seven DenseNet264 models, each designed to detect a specific artifact type: noise, zipper, positioning, banding, motion, contrast, and distortion. Similarly, as uLF MRI struggles to offer strong anatomical delineation, we enhanced challenge provided 0.064T low-field MRI scans using a custom Super-Field Network (SFNet) to generate high-quality Super-Field (SF) images, then trained an nnUNet model on these SF images with the LISA Challenge provided hippocampi segmentations. DenseNet achieved an average accuracy of 0.827 on the validation set across different artifact categories, excelling in detecting positioning and banding artifacts, with accuracies of 0.952 and 0.90, respectively. The nnUNet model trained on SF data achieved an average Dice Similarity Coefficient (DSC) of

---

A. Tapp and R. Rouhi—These authors contributed equally to the work as first authors.
M. G. Linguraru and N. Lepore—These authors contributed equally to the work as senior authors.

© The Author(s) 2025
N. Lepore and M. G. Linguraru (Eds.): LISA 2024, LNCS 15515, pp. 63–75, 2025.
https://doi.org/10.1007/978-3-031-83008-2_6

71% with the SF images-an improvement from 61% DSC using the same nnUNet; similar and significant improvements were obtained for HD, relative volume error, and average symmetric surface distance demonstrating the effectiveness of SF images in improving hippocampal segmentation accuracy. These results indicate that our automated methods effectively improve image quality assessment and hippocampal segmentation in uLF MRI, supporting the mission of the LISA 2024 Challenge and the potential adoption of portable MRI systems in resource-limited regions.

Keywords: Ultra Low Field MRI · Pediatric · Deep Learning · Quality Assurance · Segmentation · Classification

1 Introduction

Pediatric brain imaging plays a vital role in monitoring neurodevelopmental disorders, which can lead to cognitive deficits if not identified early. However, a significant challenge arises in regions with limited access to advanced medical imaging technologies such as high-field magnetic resonance imaging (MRI). Ultra-low-field (uLF) MRI, specifically the 0.064T SWOOP system (Hyperfine, Guilford, CT), has emerged as a viable alternative for such settings, providing portable, affordable imaging options, but at the cost of image quality. Within low-resource settings, the novelty of MRI as an imaging modality poses an added challenge for the healthcare staff in the area. Additionally, the poorer contrast and reduced image quality of the uLF MRI contributes to the uncertainties surrounding proper acquisition and interpretation of pediatric brain images.

The Low-field pediatric brain magnetic resonance Image Segmentation and quality Assurance (LISA) 2024 Challenge seeks to address the gap of artifact-free acquisition and key anatomical visualization by advancing automated image quality assessment and segmentation solutions tailored for pediatric populations imaged using low-field and uLF systems. The challenge focuses on two key tasks: (1) performing quality assurance (QA) of MR images affected by noise, zipper, positioning, banding, motion, contrast, and distortion, and (2) segmenting the hippocampus, a brain region crucial for cognitive functions; both tasks are performed on uLF images. Given the technical limitations of uLF MRI as well as the added challenge of infant and pediatric data, traditional image analysis methods are insufficient, necessitating the use of advanced deep learning techniques.

1.1 Quality Assurance and Image Quality

The quality of images is critical in various applications, including medical imaging, photography, and video processing. Quality degradation can occur due to factors such as noise, distortion, and inadequate contrast, which can significantly affect the interpretation and usability of images. Traditional methods for assessing image quality often rely on manual evaluation, which is time-consuming and subjective. Automated classification using deep learning models offers a promising solution to this problem. Recent advancements in automatic quality assurance have demonstrated significant enhancement in assessing MRI images. Torfeh et al. illustrated that deep learning models, such as VGG16, VGG19, and ResNet50, could effectively predict MRI image quality, achieving 80% or higher accuracy in assessing geometric distortion and spatial resolution. They noted that custom CNN models excelled in detecting low contrast, often surpassing transfer learning approaches, thereby highlighting deep learning's role in clinical MRI quality assurance [13]. Additionally, Sadri et al. introduced MRQy, an open-source tool that addresses site- and scanner-specific variations and imaging artifacts in large MR datasets, enhancing dataset consistency by reducing batch effects [10]. Samani et al. developed QC-Automator, a deep learning-based tool specifically designed for diffusion MRI quality control, achieving 98% accuracy in artifact detection, making it ideal for large-scale quality control applications [11].

Building on this foundation, Kapsner et al. demonstrated DenseNet's high effectiveness in detecting MRI artifacts, with an AUPRC of 0.921 and PPV of 0.981, showcasing its robustness despite class imbalances in the dataset [8].

Beyond quality assessment, recent studies have also highlighted deep learning's role in improving MRI image quality. Chen et al. introduced a 3D Multi-Level DenseNet combined with GAN to generate high-resolution MRI images from low-resolution inputs. This model produced sharper details and richer textures, significantly enhancing image quality and processing speed, which are vital for low-field MRI scans [3]. Kalluvila et al. (2023) explored convolutional neural networks (CNNs) for improving ULF MRI image quality, employing the Nested U-Net (U-Net++) architecture for synthetic low-field MRI resolution enhancement. Their model achieved a PSNR of 78.83 and an SSIM of 0.9551, outperforming methods like SRCNN, VDSR, and SR-GAN, thus proving CNN-based models' suitability for low-resource settings [7].

1.2 Segmentation

In the segmentation domain, recent studies have highlighted the versatility and strength of nnUNet for medical imaging challenges, specifically in brain segmentation. Donnay et al. [4] explored a pseudo-label assisted nnUNet model for 7T MRI segmentation, achieving significantly enhanced lesion detection compared to the original nnUNet. This approach demonstrated a 16% improvement in lesion DSC, indicating that transfer learning with lower field pseudo-labels can substantially enhance segmentation accuracy at high field strengths like 7T.

Another study [2] applied nnUNet for neonatal brain MRI extraction using a large multi-institutional dataset. This automated approach involved training on a substantial dataset, consisting of neonatal MRIs collected across different institutions. The study demonstrated that nnUNet could provide consistent and accurate segmentation even with diverse imaging parameters and varied MRI quality. By employing iterative model development with human-in-the-loop corrections, the approach achieved a DSC of 0.955, outperforming other traditional and automated methods for brain extraction. This finding underlines the importance of using nnUNet for robust segmentation tasks, particularly when dealing with variations common in neonatal and low-field MRI data.

1.3 Contributions

In this paper, we propose one QA method and one segmentation method to tackle the LISA 2024 Challenge. First, for the QA task, we leverage DenseNet [5], a convolutional neural network (CNN) architecture known for its strong feature propagation capabilities and ability to classify images based on different grades of quality degradation. Notably, we train 7 specialized DenseNets, and each one is tailored to classify if a specific artifact is present in the pediatric uLF MRI dataset. Second, for hippocampal segmentation, we adopt nnUNet [6], a self-configuring deep-learning framework specifically designed for medical image segmentation. However, we do not utilize the original, LISA-challenge-provided uLF images in the nnUNet training process. Instead, we leverage an enhancement approach through a custom preprocessing pipeline that improves the anatomical visibility of uLF structures, like the hippocampus. The method, referred to as a Super-Field (SF) approach, is based on our previous work [12]. To super-field an image, we use the Super-Field Network (SFNet), a custom swinUNETRv2 with generative adversarial network components. The network generates images comparable to HF MRIs from original uLF MRIs. During segmentation network training, we utilized the nnUNet model but provided SF images, instead of uLF images, as the training data. Utilizing SF images ultimately provided more accurate hippocampal segmentations than relying solely on uLF images. Our tailored enhancement approach underscores the value of targeted improvements for the unique challenges of pediatric uLF MRI, further contributing to advancements in both quality assurance and image enhancement.

2 Methods

2.1 Dataset

For QA, the training dataset includes 432 pediatric brain MRI scans at 0.064T from 144 individuals. Each individual's brain was orthogonally acquired, in three directions along each anatomical plane: axial, coronal, and sagittal using a Hyperfine SWOOP using a magnetic field strength of 0.064T, with a sequence type of spin echo, TR 1.5 s, TE 5ms, TI 400ms. Quality assessment scores for seven artifact domains were manually determined by image analysis experts and provided with the training dataset in an accompanying CSV file. Figure 1 presents examples for each artifact category. The validation and hidden testing cohorts consisted of 42 images from 14 subjects and 27 images from 9 subjects, respectively.

For the segmentation task, 79 pairs of super-resolution reconstructed (SRR) low-field MRI scans were provided. The provided SRR MRIs are the combined version of the three orthogonally acquired images mentioned above. These SRR

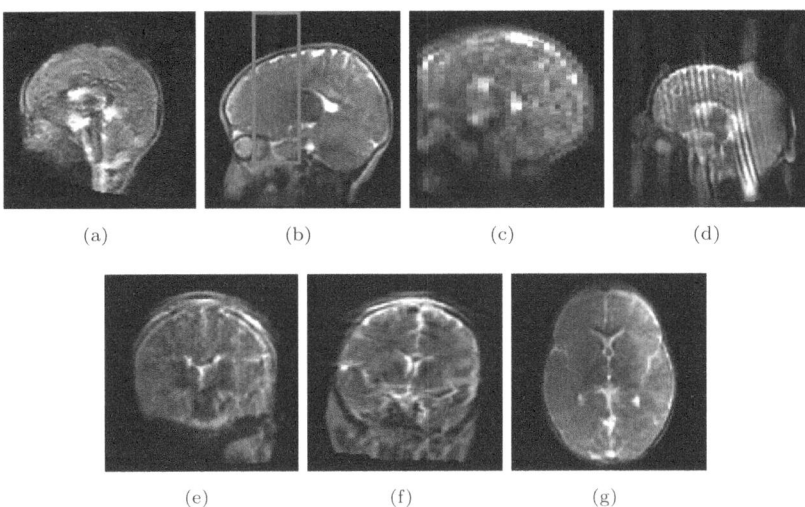

Fig. 1. Artifact examples in low-field MR images: (a) noise, (b) zipper (specified in the red box), (c) positioning, (d) banding, (e) motion, (f) distortion and (g) contrast. (Color figure online)

MRIs have isotropic spacing and are paired with bilateral hippocampal segmentations derived directly from the HF MRIs of the same patient. Segmentations were manually performed on HF T2 MRI images (3T or 1.5T), with corrections reviewed by a board-certified pediatric neuroradiologist. Both the uLF MRIs and the hippocampal segmentations were aligned with the corresponding subject's high-field MRI space. The validation and hidden testing cohorts consisted of 12 and 9 images, respectively.

2.2 Quality Assurance with DenseNet

The quality assurance task utilizes the LISA dataset, which contains pediatric low-field MRIs with various artifacts across seven domains: noise, zipper, positioning, banding, motion, contrast, and distortion. Each image is scored on a scale from 0 to 2: a score of 0 indicates no presence of the artifact, a score of 1 indicates the artifact's presence with some interference that may compromise the distinction between neighboring subcortical structures, and a score of 2 indicates that the artifact significantly compromises the distinction between these structures. DenseNet [5] was chosen for its ability to learn rich feature hierarchies while maintaining gradient flow between layers. DenseNet's architecture allows for densely connected blocks, which mitigate the vanishing gradient problem and promote feature reuse. DenseNet is offered in several variants, including DenseNet121, DenseNet169, DenseNet201, and DenseNet264, which are primarily distinguished by their depth and layer count. For our purposes, we selected the latest version, DenseNet264, as implemented in MONAI 0.9.0 [1]. Each layer receives input from all previous layers, making it well-suited for tasks that require robust extraction of low-level and high-level features. The architecture for this task includes multiple DenseNet264 networks, each network is trained to predict a single artifact. Thus, 7 total networks were trained. Orthogonal images were used as inputs, and the network's output was a prediction of one of the grades, 0, 1, or 2, corresponding to the specific artifact the network was designed to assess. During training, we used the weighted F1-score [9] to validate the classification performance at the end of each epoch, as our dataset was imbalanced. The weighted F1-score calculates the F1-score for each class (grade) and averages them, weighted by the number of true instances in each class. This ensures that larger classes have more influence, addressing imbalances by preventing dominant classes from overshadowing others, resulting in a fairer evaluation across all classes. During inference, each of the 7 networks was deployed across a single case to yield a single artifact prediction (0, 1 or 2). The model was implemented using PyTorch, trained on 80% of the LISA dataset, with the remaining 20% used for validation. We employed the randomly cropped images with a size of $256 \times 256 \times 256$, an Adam optimizer with a learning rate of 10^{-5}, cross-entropy loss, a batch size of 2, and trained the model for 100 epochs. All models were trained using an NVIDIA RTX A6000 GFORCE 64Go GPU.

2.3 Hippocampal Segmentation with nnUNet

For the hippocampal segmentation task, nnUNet was selected due to its adaptability to a wide variety of segmentation tasks [6]. nnUNet automatically adjusts hyperparameters such as learning rates, patch sizes, and batch sizes based on the input data. To further improve segmentation performance, we applied a technique to enhance the provided low-field MRIs, transforming them into super-field images (Fig. 2). These SF images were generated using a custom Super-Field Network, SFNet, based on the model presented in [12]. SFNet is a specialized swinUNETRv2 network with generative adversarial components, designed to produce high-quality MRIs from low-field data. It was originally trained on a private cohort of 90 paired low-field (LF) and high-field (HF) MR images of infant brains. SFNet employs a learned discriminator to evaluate output images on a global scale, resulting in more realistic images with reduced anti-aliasing effects due to the combined use of adversarial loss, structural similarity index measure (SSIM) loss, and perceptual component loss. When compared to existing methods, SFNet has been shown to improve segmentation accuracy for gross brain structures (white matter, gray matter, and cerebrospinal fluid); more details on SFNet can be found in [12]. For this challenge we trained the nnUNet model using only the SF-enhanced images derived from uLF images. The nnUNet was trained with the full resolution configuration, as described in [6], on the provided hippocampal segmentation with cross-entropy Dice similarity coefficient as the primary loss function. To validate the efficacy of anatomical enhancement seen in the SF images, the challenge provided uLF images were used to train a second nnUNet. Both networks were trained on an A5000 GPU, utilizing the same 66 training images and 13 validation images in a 5-fold cross-validation process. To assess significant differences between the network segmentation performance, we utilize the Wilcoxon signed-rank test.

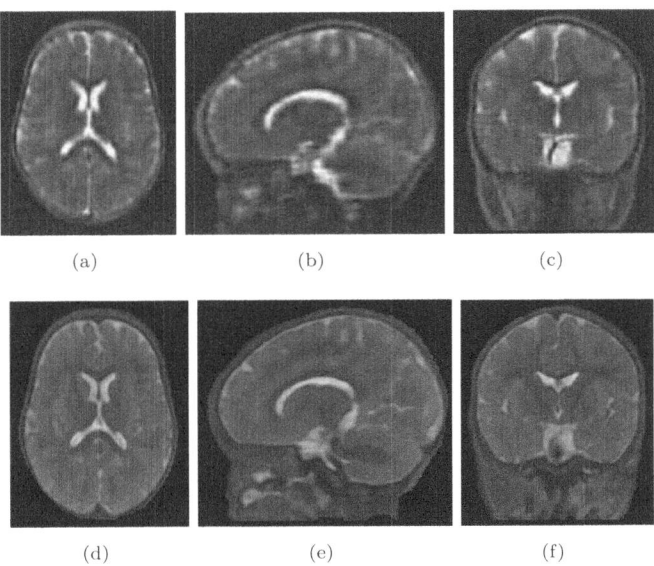

Fig. 2. Super-fielding the LISA challenge data. The challenge provided ultra-low-field MR images: views of the (a) axial, (b) sagittal, (c) coronal planes, and the enhanced version of these images in (d) axial, (e) sagittal, and (f) coronal planes.

3 Results

This section presents the results for each task, divided into two parts: quality assurance (QA) and segmentation. The QA part reviews artifact assessment performance, while the segmentation part focuses on anatomical accuracy. Together, these results highlight the model's performance across both tasks.

3.1 Quality Assurance with DenseNet

Figure 3 displays examples of validation cohort images that were accurately classified and misclassified by the trained models. Given the predicted vectors from each of the 7 networks, including the grades of each artifact for validation images, and the corresponding ground truth vector in the provided CSV file, we calculated various metrics for each artifact category. As shown in Table 1, the DenseNet models achieved average accuracies of 0.827 on the validation set, respectively. We also presented the results for both micro and macro cases across different metrics [9]. Table 2 shows the results based on the weighted case for different categories respectively for the validation. The highest metric values were achieved in the positioning and banding categories, while the lowest

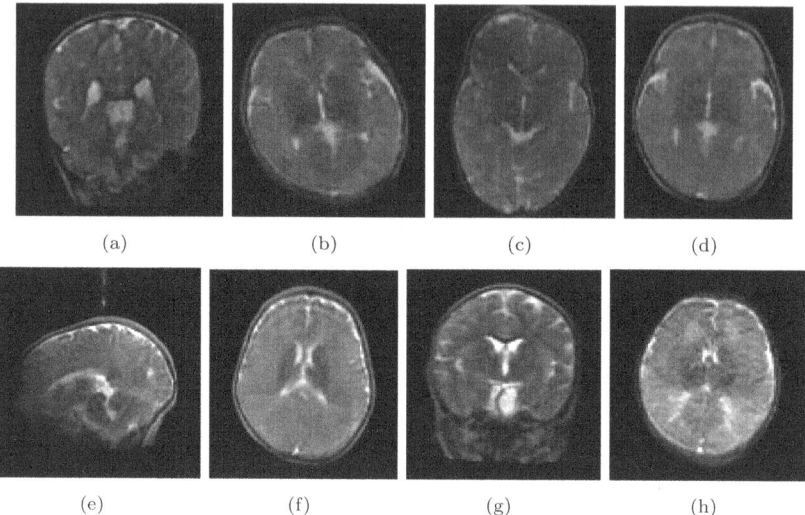

Fig. 3. The figure presents sample images from the validation cohort. The top row shows images that were correctly classified in the QA with a grade of 1 in the following categories: (a) positioning, (b) motion, (c) distortion, and (d) contrast. The bottom row displays images that were misclassified in the QA with a grade of 0 for the same categories: (e) positioning, (f) motion, (g) distortion, and (h) contrast.

Table 1. QA results for the validation cohort, presented with values averaged across micro, macro, and weighted metrics for DenseNet264 models trained on each artifact category.

Metrics	Precision	Recall	F1-score	F2-score	Accuracy
Micro	0.827	0.827	0.827	0.827	0.827
Macro	0.510	0.525	0.506	0.514	0.827
Weighted	0.829	0.827	0.825	0.825	0.827

were observed in categories like zipper, contrast and noise (see last column of Table 2), reflecting the inherent challenges in identifying zipper, contrast and Noise artifacts related to low-filed pediatric brain imaging.

3.2 Hippocampal Segmentation with nnUNet

The nnUNet model trained on Super-Field (SF) images demonstrated significant improvements in hippocampal segmentation accuracy on the validation set of 12 cases. Inference was successfully performed on these cases after enhancing the challenge provided uLF MRI scans using SFNet.

As shown in Table 3, overall, the right hippocampus segmentation outperformed the left, with an average DSC of 0.74 ± 0.13 for the right hippocampus compared to 0.68 ± 0.24 for the left. The left hippocampus had an average HD of

Table 2. QA validation cohort results, evaluated using different metrics for DenseNet264 models trained on each artifact category.

Category	Precision	Recall	F1-score	F2-score	Accuracy
Noise	0.722	0.762	0.741	0.754	0.762
Motion	0.823	0.81	0.816	0.812	0.810
Contrast	0.827	0.762	0.789	0.771	0.762
Positioning	0.940	0.952	0.944	0.948	0.952
Banding	0.951	0.905	0.927	0.914	0.905
Zipper	0.690	0.714	0.701	0.709	0.714
Distortion	0.852	0.881	0.858	0.871	0.881

Table 3. Hippocampus (Hipp) segmentation results on the uLF validation cohort. The first block of results displays metrics for nnUnet predictions on Super-Field (SF) images, which are uLF-enhanced images created using an existing, pretrained network by Tapp et al. [12]. The SF nnUnet was trained using SF images and LISA-provided ground truth segmentations. The second block of results displays metrics for the nnUNet predictions on the original uLF validation data; this nnUNet was also trained on original uLF data, not SF data. * indicates SF nnUnet performance was significantly ($p < 0.05$) better than performance by the uLF-trained nnUNet.

Model	Metrics	DSC	HD	HD-95	ASSD	RVE
SF nnUNet	Left Hipp	0.68 ± 0.24*	7.93 ± 15.93*	2.21 ± 1.90*	0.85 ± 1.13*	0.19 ± 0.14
SF nnUNet	Right Hipp	0.74 ± 0.13	3.17 ± 0.71*	1.77 ± 0.70*	0.53 ± 0.33*	0.13 ± 0.11
SF nnUNet	Average	0.71 ± 0.17	5.55 ± 7.96*	1.99 ± 1.12*	0.69 ± 0.65*	0.16 ± 0.11
uLF nnUNet	Left Hipp	0.57 ± 0.29	14.82 ± 20.24	6.18 ± 12.38	4.10 ± 10.55	0.18 ± 0.15
uLF nnUNet	Right Hipp	0.65 ± 0.25	12.28 ± 16.50	10.67 ± 16.07	3.96 ± 10.38	0.14 ± 0.13
uLF nnUNet	Average	0.61 ± 0.27	13.55 ± 15.77	8.43 ± 13.08	4.03 ± 10.45	0.16 ± 0.11

7.93 ± 15.93 mm and an HD95 of 2.21 ± 1.90 mm, indicating variability due to a segmentation issue. In contrast, the right hippocampus showed more consistent results with an average HD of 3.17 ± 0.71 mm and an HD95 of 1.77 ± 0.70 mm. Average HD was 5.55 ± 7.96 mm, and average HD95 was 1.99 ± 1.12 mm. The Average Symmetric Surface Distance (ASSD) and Relative Volume Error (RVE) also indicated better performance for the right hippocampus. The left hippocampus had an ASSD of 0.85 ± 1.13 mm and an RVE of 0.19 ± 0.14, whereas the right hippocampus achieved an ASSD of 0.53 ± 0.33 mm and an RVE of 0.13 ± 0.11. The combined averages were an ASSD of 0.69 ± 0.65 mm and an RVE of 0.16 ± 0.11. Comparatively, the nnUNet model trained on the original uLF images without SF enhancement achieved a lower average DSC of 0.61, underscoring the efficacy of using SF images for training. The improvements in HD, HD95, ASSD, and RVE metrics corroborate the enhanced segmentation performance afforded by the SF images. For each comparison, the Wilcoxon ranked sign test using a p-value with a significance threshold of $p < 0.05$ was used to identify sta-

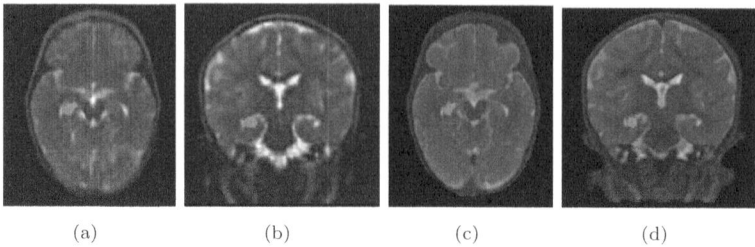

(a)　　　　　　(b)　　　　　　(c)　　　　　　(d)

Fig. 4. Example of the left (red) and right (green) hippocampi segmentations overlayed ultra-low-field MR images in (a) axial and (b) coronal views and for SF images in the (c) axial, (d) coronal views. Right and left hippocampi segmentation dice scores are 68% and 71% for ultra-low-field cases, respectively, 78% and 83% for super-field predictions, respectively. (Color figure online)

tistically meaningful differences between the two methods across the evaluated metrics. Statistically significant improvements included left hippocampus DSC, all HD values, left and average HD95 and all ASSD. These results demonstrate that enhancing uLF MRI scans using SFNet significantly improves hippocampal segmentation accuracy, particularly for the right hippocampus.

The disparity in left hippocampus performance highlights areas for further refinement. The success in the highlighted case (Fig. 4) exemplifies the potential of the proposed method in achieving high-precision segmentation, which is crucial for clinical applications in pediatric neuroimaging.

4　Discussion

This study demonstrates the potential of using advanced deep learning methods, specifically DenseNet and nnUNet, for analyzing low-field pediatric MR images. The results demonstrate the effectiveness of using DenseNet for the automated classification of image quality degradation. By training separate models for each category, we ensured that the models specialized in recognizing the nuances specific to noise, distortion, contrast, banding, zipper, motion, and positioning, improving the overall classification performance. This tailored approach not only enhances accuracy but also allows for more accurate identification of subtle variations in image quality that might otherwise be overlooked in a generalized model. This approach can be extended to other types of image impairments, offering a scalable solution for automated quality assurance across diverse imaging datasets. However, additional data augmentation techniques may be necessary to enhance model performance, particularly for more complex artifacts such as motion and contrast distortions, where variability is higher.

Implementing advanced augmentation techniques, such as simulated artifact injection or synthetic data generation, could further diversify the training set and improve the model's robustness in real-world scenarios. In training, we used the weighted F1-score for validation evaluation to address the issue of data imbalance, ensuring that underrepresented categories, motion and contrast distortions, were accounted for in the model's performance. The F1-score is particularly useful because it incorporates both precision and recall into a single metric, balancing the trade-offs between false positives and false negatives. Additionally, alternative metrics could also be considered and might yield improved training performance.

A key finding of this study is the significant improvement in hippocampal segmentation accuracy through SFNet-enhanced uLF MRI scans. The nnUNet model trained on SF images achieved an average DSC of 71% compared to 61% achieved by the model trained on challenge-provided uLF images. The 10% improvement in segmentation accuracy similarly reflected a decrease in surface and volume discrepancy. The metric improvements can be attributed to the increased anatomical visualization and enhanced contrast provided by the SF images. The SFNet effectively transforms uLF MRI scans into higher-quality images with enhanced structural details, allowing the nnUNet model to better distinguish anatomical boundaries. In low-field MRI, the inherent low signal-to-noise ratio and poor contrast can obscure fine anatomical structures like the hippocampus. By enhancing image quality, SFNet mitigates these limitations, enabling more accurate and reliable segmentation. However, challenges remain in achieving optimal segmentation performance in areas of the right hippocampus. The HD and the HD-95 further reflected the disparity between the left and right hippocampus segmentation performance. Future improvements could include incorporating additional data augmentation for greater model robustness, refining SFNet to handle more image variations, and integrating multi-modality data to enhance anatomical contrast. Exploring transfer learning or domain adaptation could further improve segmentation in challenging cases. The combination of SFNet and nnUNet offers a promising approach to enhancing hippocampal segmentation in uLF MRI scans.

By leveraging the enhanced anatomical visualization and contrast provided by SF images, deep learning models can achieve significantly higher accuracy, which is crucial for clinical applications in pediatric neuroimaging-especially in resource-limited settings where high-field MRI is not available. Significant improvements in segmentation metrics confirm SFNet's effectiveness in addressing low-field MRI challenges and highlight its potential to enhance diagnostics in under-resourced regions.

5 Conclusion

This paper presents two approaches addressing the LISA 2024 Challenge tasks on QA and hippocampal segmentation for pediatric low-field MRI. Portable uLF, 0.064T MRI offers a valuable solution for resource-limited regions, but the modality's novelty requires automated methods to ensure image quality and anatomical segmentation. We used DenseNet for QA, achieving 0.827 accuracy in detecting artifacts like positioning and banding. For hippocampal segmentation, we trained nnUNet using the enhanced SF images, achieving a Dice Similarity

Coefficient (DSC) of 71% without the use of specialized augmentations or complex preprocessing steps and relyed solely on image enhancement and the default nnUNet configuration. These results demonstrate the potential of DenseNet and nnUNet for uLF MRI tasks, though further optimization is needed. By refining these methods, we aim to advance pediatric neuroimaging and support portable MRI adoption in low-resource settings.

Acknowledgments. This work was supported by the Bill & Melinda Gates Foundation under investments INV-005798, INV-047887, INV-018164, INV-004939, and INV-023509, as well as the Wellcome Leap 1kD programme (The First 1000 d; 222076/Z/20/Z).

Disclosure of Interests. The authors have no competing interests to declare that are relevant to the content of this article.

References

1. https://doi.org/10.5281/zenodo.6639453
2. Chen, J.V., et al.: Automated neonatal NNU-net brain MRI extractor trained on a large multi-institutional dataset. Sci. Rep. **14**(1), 4583 (2024)
3. Chen, Y., Christodoulou, A.G., Zhou, Z., Shi, F., Xie, Y., Li, D.: MRI super-resolution with GAN and 3D multi-level densenet: smaller, faster, and better. arXiv preprint arXiv:2003.01217 (2020)
4. Donnay, C., et al.: Pseudo-label assisted nnU-net enables automatic segmentation of 7t MRI from a single acquisition. Front. Neuroimaging **2**, 1252261 (2023)
5. Huang, G., Liu, Z., Van Der Maaten, L., Weinberger, K.Q.: Densely connected convolutional networks. In: Proceedings of the IEEE Conference on Computer Vision and Pattern Recognition, pp. 4700–4708 (2017)
6. Isensee, F., Jaeger, P.F., Kohl, S.A., Petersen, J., Maier-Hein, K.H.: nnU-net: a self-configuring method for deep learning-based biomedical image segmentation. Nat. Methods **18**(2), 203–211 (2021)
7. Kalluvila, A., Koonjoo, N., Bhutto, D., Rockenbach, M., Rosen, M.S.: Synthetic low-field MRI super-resolution via nested u-net architecture. arXiv preprint arXiv:2211.15047 (2022)
8. Kapsner, L.A., et al.: Image quality assessment using deep learning in high b-value diffusion-weighted breast MRI. Sci. Rep. **13**(1), 10549 (2023)
9. Pedregosa, F., et al.: Scikit-learn: machine learning in python. J. Mach. Learn. Res. **12**, 2825–2830 (2011)
10. Sadri, A.R., et al.: Mrqy-an open-source tool for quality control of MR imaging data. Med. Phys. **47**(12), 6029–6038 (2020)
11. Samani, Z.R., Alappatt, J.A., Parker, D., Ismail, A.A.O., Verma, R.: Qc-automator: deep learning-based automated quality control for diffusion MR images. Front. Neurosci. **13**, 1456 (2020)
12. Tapp, A., et al.: Super-field MRI synthesis for infant brains enhanced by dual channel latent diffusion. In: Proceedings of Medical Image Computing and Computer Assisted Intervention – MICCAI 2024, vol. LNCS 15003. Springer Nature Switzerland (2024)
13. Torfeh, T., Aouadi, S., Yoganathan, S., Paloor, S., Hammoud, R., Al-Hammadi, N.: Deep learning approaches for automatic quality assurance of magnetic resonance images using ACR phantom. BMC Med. Imaging **23**(1), 197 (2023)

Open Access This chapter is licensed under the terms of the Creative Commons Attribution 4.0 International License (http://creativecommons.org/licenses/by/4.0/), which permits use, sharing, adaptation, distribution and reproduction in any medium or format, as long as you give appropriate credit to the original author(s) and the source, provide a link to the Creative Commons license and indicate if changes were made.

The images or other third party material in this chapter are included in the chapter's Creative Commons license, unless indicated otherwise in a credit line to the material. If material is not included in the chapter's Creative Commons license and your intended use is not permitted by statutory regulation or exceeds the permitted use, you will need to obtain permission directly from the copyright holder.

Author Index

C
Cai, Rongqing 3
Chen, Geng 3
Chen, Zhaolin 15
Choi, Jinwha 53

D
Deoni, Sean 63
Dinsdale, Nicola K 41

J
Jiang, Haotian 3

K
Kim, Hyunwook 53

L
Lepore, Natasha 63
Li, Jingyu 28
Linguraru, Marius George 63
Lyu, Mengye 28

O
On, Sungchul 53

P
Park, Joon hyung 53
Peiris, Himashi 15

R
Rouhi, Rahimeh 63
Ryu, Seiyoung 53

S
Seo, Jinew 53
Sundaresan, Vaanathi 41

T
Tanedo, Jeffrey 63
Tapp, Austin 63

W
Wang, Xi 28
Wang, Yuwan 28

Z
Zanjal, Shreyash 63
Zhou, Weichen 28
Zhu, Yueyue 3

The manufacturer's authorised representative in the EU is Springer Nature Customer Service Centre GmbH, Europaplatz 3, 69115 Heidelberg, Germany. If you have any concerns regarding our products, please contact ProductSafety@springernature.com

Printed and bound by CPI Group (UK) Ltd, Croydon, CR0 4YY

26/03/2026

02078935-0014